COMPRESSOR APPLICATION ENGINEERING

VOLUME 2

....

Drivers for Rotating Equipment

Volume 1 Compression Equipment

Chapter 1   Thermodynamic Laws and Formulas
Chapter 2   Compressor Efficiency
Chapter 3   Positive Displacement Compressors—Common Characteristics
Chapter 4   Blowers with Rotating Lobes (Roots Blowers)
Chapter 5   Sliding-Vane Compressors
Chapter 6   Screw-Type Rotary Compressors
Chapter 7   Reciprocating Compressors
Chapter 8   Characteristics and Operating Principles of Turbocompressors
Chapter 9   Components of Turbocompressors
Chapter 10 Centrifugal Compressors
Chapter 11 Axial Compressors
Appendixes

Volume 2 Drivers for Rotating Equipment

Chapter 1   Electric Motors
Chapter 2   General Characteristics of Heat Engines
Chapter 3   Gas Engines
Chapter 4   Gas Turbines
Chapter 5   Free-Piston Gas Generators
Chapter 6   Steam Turbines—The Rankine Cycle
Chapter 7   Couplings

COMPRESSOR APPLICATION ENGINEERING

VOURME 2

Drivers for Rotating Equipment

Pierre Pichot

Translated from the French by
Ryle Miller and Ethel B. Miller

Gulf Publishing Company
Book Division
Houston, London, Paris, Tokyo

COMPRESSOR APPLICATION ENGINEERING

VOLUME 2

Drivers for Rotating Equipment

Copyright © 1986 by Gulf Publishing Company, Houston, Texas. All rights reserved. Printed in the United States of America. This book, or parts thereof, may not be reproduced in any form without permission of the publishers.

This book represents the personal experience and opinions of the author, who is entirely responsible for its content.

Library of Congress Cataloging-in-Publication Data
Pichot, Pierre.
  Compressor application engineering.

  Includes indexes.
  Contents: v. 1. Compression equipment—
v. 2. Drivers.
  1. Compressors. 2. Fans (Machinery) 3. Electric motors. 4. Heat-engines. I. Title.
TJ990.P5 1986   621.5   86-7587
ISBN 0-87201-705-2 (v. 1)
ISBN 0-87201-706-0 (v. 2)

Contents

Preface, vii

Notation, ix

Chapter 1   Electric Motors,   1
Fixed-Speed Alternating Current Motors. Variable-Speed Alternating Current Motors.

Chapter 2   General Characteristics of Heat Engines,   16
Thermal Efficiency of a Heat Engine—Specific Consumption. Theoretical Thermodynamic Efficiency. Shape Efficiencies for Thermodynamic Cycles. Combustion Efficiency. Mechanical Efficiency. Indicated Efficiency. Approximate Efficiencies for Reciprocating Internal Combustion Engines. Air Pollution by Heat Engines. Intake Air Filters and Silencers.

Chapter 3   Gas Engines,   50
Characteristics of Gas Engines. Thermodynamic Cycle. Ignition. Piston Rings. Supercharging: Influence of Altitude and Supercharging as a Function of Ambient Temperature. Weights, Floor Space, and Rotating Speeds. Gas Engine Operation.

Chapter 4   Gas Turbines,   108
Technical Description of a Gas Turbine. Thermodynamic Cycles. Design Criteria for Gas Turbines. Gas Turbine Design. Reducing Nitrogen Oxides in the Exhaust. Existing Gas Turbines. Operating Characteristics and Use. Corrections for Other than Standard Operating Conditions. Operation and Control—Suction and Exhaust Silencers and Air Filters.

Chapter 5   Free-Piston Gas Generators,   184
Operation of a Free-Piston Gas Generator. Operating Characteristics. Existing Free-Piston Gas Generators.

v

Chapter 6   Steam Turbines—The Rankine Cycle,   188
Thermal Efficiency of the Rankine Cycle. Choice of Fluid. Steam Generation in the Steam-Gas Turbine Combined Cycle. Calculating Performance for a Gas Turbine with a Waste Heat Boiler and Steam Turbine. Feeding the Waste Heat Boiler. Characteristics of Steam Turbines. Impulse Turbines. Reaction Turbines. Features of Steam Turbine Operation.

Chapter 7   Couplings,   233
Direct Couplings. Speed-Reducing and Speed-Increasing Couplings. Variable-Speed Couplings.

Index,   245

Preface

Drivers for rotating machines do not normally come into direct contact with the feed and product materials of a processing plant. Consequently, drivers are often thought of as secondary in importance to the compressor, pump, conveyor, mixer, grinder, centrifuge, or other machine to which the driver supplies power. However, the mere fact that all machines operate at something less than 100% efficiency means that drivers must handle more energy than the machines they serve. Because energy is a principal raw material of many manufacturing processes, drivers thus have a significance greater than the rotating machinery. This applies in particular to gas pipeline transmission and electric power generation. Because compressors and blowers are among the foremost contemporary power consumers, drivers are particularly important wherever compressors and blowers are used.

Consequently, this book, which is a companion to *Compression Equipment,* Volume 1 of the two-volume series *Compressor Application Engineering,* emphasizes drivers for electrical generating stations and for compressors used in gas transmission systems. Because two of the most important contemporary drivers, gas engines and gas turbines, share the general characteristics of heat engines, those characteristics are discussed separately. And because most drivers need a coupling, couplings are also discussed.

Based on the relative cost of the different available forms of energy, one can select the driver best suited to the characteristics of the appropriate driven machine. Thus users of this book can refer to Volume 1 to arrive at an optimum combination to suit the specific purpose.

Pierre Pichot

Notation

A	constant, $A = 1/J$
a	area under a curve
C	torque
cal	calories
metric HP	metric horsepower = 0.0986 hp
Cp	molal specific heat at constant pressure
Cv	molal specific heat at constant volume
c_p	specific heat at constant pressure for 1 kg
c_v	specific heat at constant volume for 1 kg
d	specific gravity of a gas with respect to air = $M/28.966$
ϵ	proportionality factor
g	gram; acceleration due to gravity
γ	c_p/c_v
h	hours
H	head; pressure expressed either as meters of elevation of flowing fluid or as joule
i	enthalpy
I	mass flow per second
J	4,184 joule/kcal
j	joule
kg	kilogram mass
kgf	kilogram force (weight), 1 kgf = 9.81 Newton
k	kilo
l	length
$m^3(n)$	cubic meter at normal conditions (273 K, 760 mm Hg)
m	meter
mbar	millibar
mm	millimeter
M	molecular mass
N	Newton; rotation speed, rpm
w	specific weight (force); $= \rho \cdot g$
π	$= 3.146$

p	pressure
Q	heat exchange
ζ	pressure drop
ρ	density (kg/m^3) or specific mass; efficiency
r	perfect gas constant in terms of mass = R/M
R	perfect gas constant in terms of molecules = rM
S	entropy; area under a curve; Siemens electrical conductance = mho
$\mathfrak{J}$	work energy
$\mathfrak{J}_e$	work exerted by outside forces; external work
T	absolute temperature, K
U	internal energy
v	specific volume for 1 kg; subscript volume
V	velocity; volume
W	kinetic energy; power; watt
Z	compressibility factor of a gas
ω	angular velocity

Subscripts

0, 1, 2, 3, ... n	state; stage
ave	average
ad	adiabatic
c	critical value at which vapors and liquids follow the same progressions
e	Neperian or natural logarithm
ext	external
i	internal
iso	isothermal
m	average
p	pressure; constant pressure
pol	polytropic
r	reduced value relative to the critical value
v	volume
bar	= 10^5 pascal, 1 atm = 1,013 bar = 1,033 kgf/cm^2
°C	degree Celsius; degree Kelvin, K = 273 + °C

Unit Systems

SI	international metric system (m, kg, sec, pascal, joule, K, Newton)
m kgf S	former metric system sometimes still used (m, kgf, sec, kgf/m^2, kgf·m, K, mass 1 = 1 kgf/g)

Chapter 1
Electric
Motors

Electric motors convert electrical energy to mechanical energy through the force exerted between a magnetic field and an electric-current-carrying conductor. The typical motor is cylindrical, with the magnetic field imposed by an axial rotating core, called the rotor, and with the current-carrying conductor, called the stator, imbedded in the cylindrical walls. If the magnetic field is excited by wires wound on the rotor, the motor is called "wound rotor." If the field is excited by a cage of bars secured at their ends to two rings parallel to the axis, the rotor is called "squirrel cage."

Because of the reversal in voltage as a conductor moves from one to the opposite pole of a magnetic field, the sinusoidally varying voltages of commercial alternating current are suited to electric motors. The speed of some motors is directly related to the frequency of the voltage and the number of magnetic poles in the motor. Thus

$$N = 120 \, f/p$$

where N = speed in revolutions per minute
       f = frequency of the current alternations, cycles per second
       p = number of magnetic poles

However, an impedance exists in alternating-current motors and circuits, such that the voltage, V, and the amperage, I, do not reach their sinusoidal peaks at the same instant. Thus the apparent power, S, equal to the product of voltage and amperage (S = VI) will not correspond to the delivered power, P. This relation is expressed as:

$$W = S \cos \theta = VI \cos \theta = I^2 R$$

where   R = resistance
       $\cos \theta$ = power factor; = R/Z

where   Z = impedance

Since the phase angle θ between voltage and current can vary from zero to 90°, the power factor can vary from 0 to 1.0, depending on the character of the circuit. The energy over a complete alternating-current circuit is returned to the circuit without loss. However, it is frequently desirable to balance a load with a high impedance (lagging phase angle θ) with a load of high capacitance (advanced phase angle θ).

FIXED-SPEED ALTERNATING CURRENT MOTORS

Nonsynchronous Motors

These motors can be either wound or squirrel-cage. They are simple and rugged; they offer a high startup torque on the order of twice normal. However, they have a generally poor power factor that can be as low as 0.1–0.2 at small or negligible loads. Generally used at a nominal torque equal to 50% of their maximum, their operation is stable and their speed is not sensitive to a variation of load. Based on rotor resistance alone, the startup current can be assumed as 4–10 times the nominal current.

Nonsynchronous motors are commonly started up by one of four methods as follows:

1. The stator's three phases are star-coupled, so that the startup current to the stator is reduced by a factor of $(3)^{1/2}$ of the relative applied load. This means that the startup torque is reduced by a factor of 3, because the torque is proportional to the square of the current. Thus this method is suitable for motors started up under low load.
2. The stator's three phases have the current reduced by a rheostat, which produces the same results as the star-coupled method, except that the reduction of current is adjustable to any desired value. This method is technically poor, because it involves large energy losses during startup, and the startup torque is reduced excessively.
3. An automatic startup transformer reduces torque and current proportionally to the square of the voltage ratio. This method is more efficient than the preceding, but more expensive.
4. A three-phase rheostat in series reduces current to the rotor. Because the maximum torque of a nonsynchronous motor is independent of the rotor's resistance, it is thus possible to reduce the startup current through the rotor and have a startup torque as high as desired. This method permits startup under load, while reducing the current as the speed increases and the resistance changes. Large motors often use liquid rheostats of electrodes progressively immersed in a bath of electrolyte. This startup method is most commonly used; but it applies only to motors with wound rotors.

Also, motors with less than 40 kW power can be equipped with a centrifugal automatic switch, which performs the function of the startup rheostat by progressively eliminating internal resistance through short circuiting the centrifugal contacts.

Special Startup Rotors

The rotor of a double-cage motor has two sets of bars, one surrounding the other. The external cage, which is the closest to the resistant air gap between the rotor and stator (Figure 1-1) functions alone at startup and gives a high startup torque of twice normal. The internal cage, which has low resistance and a high inductance loss, produces the torque for normal operation. Because of the high-frequency variation of the rotor flux during startup, the current in the internal cage is greatly reduced due to Faraday's law:

$$\frac{d}{dt}(\Phi + \lambda i) + ri = 0$$

where Φ = flux in the air gap
 λ = (large) inductance loss in the internal cage
 r = (small) resistance of the internal cage
 i = current to the internal cage

The current during startup is 2–6 times the normal current.

Also, it is possible to achieve these effects by constructing rotors with deep slots and high conductor bars (Figure 1-1). During startup, when

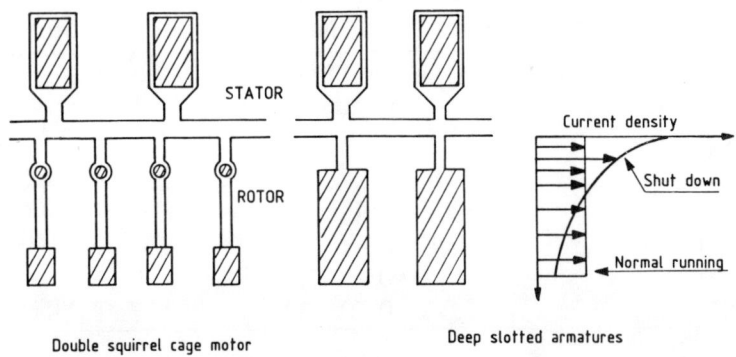

Figure 1-1. Double-squirrel-cage or deep-slotted rotor.

the frequency of current variation is the highest, the current tends to localize in the exterior part of the conductors, whereas all of the conductor is active during normal operation. The result is a variation of the apparent resistance of the rotor ranging from 1 to 5, depending on the speed.

Rotors specially designed for startup in these ways permit direct on-line startup at a moderate current and a high torque.

Efficiencies and Power Factors

Typical efficiencies for nonsynchronous motors are shown in Table 1-1. For low speeds, these efficiencies are reduced by 0.5%–2%. Efficiencies for high and low voltages are not distinguished, because they are essentially the same for motors rated over 100 kW. Below this, motors fed with high voltage have an efficiency reduced by 1%–3%.

Power factors typical of these motors are shown in Table 1-2.

Table 1-1
Typical Efficiencies for Nonsynchronous Motors

Power (kW)	4/4 Load	3/4 Load	1/2 Load
25–50	89	88.5	86.5
75–100	91	90	88
150–200	92	91.5	89.5
More than 200	93.5	93	91

Table 1-2
Typical Power Factors for Nonsynchronous Motors

Power (kW)	4/4 and 3/4 Load	1/2 Load
3,000	0.90	0.87
1,500	0.87	0.84
1,000	0.85	0.82
750	0.81	0.78
600	0.75	0.72
500	0.70	0.67

Synchronous Motors

Synchronous motors have pronounced magnetic poles; at 60 Hertz alternating-current cycle frequency, their speeds are theoretically 3,600, 1,800, 1,200, or 900 rpm, depending on whether they have 2, 4, 6, or 8 poles, respectively; and they can furnish a torque only at their speed of synchronization. Consequently, their startup torque is nil, but they offer the advantage of being able to operate at a power factor of cos $\theta = 1$, or even higher.

The most common method for bringing these rotors up to synchronous speed consists in starting up the motor nonsynchronously. Each pole shoe carries a buffer winding of little bronze bars connected to each other at the ends by copper or brass short-circuit rings. This winding behaves as a squirrel cage and produces an asynchronous torque high enough to turn over the rotor, providing the moment of inertia and the startup torque are not too high. During the whole startup period, the field conductors have no role whatsoever and are either open or blocked by a resistance. It is sufficient to feed these conductors at near synchronous speed, and the motor then engages itself.

Because of the negligible resistance, the startup current would be able to reach 5–10 times normal, if under full voltage; and to prevent such a current overload, an autotransformer is required to reduce voltage and, incidentally, the startup torque.

The buffer-winding bars of large motors can be replaced by three-phased windings connected to a liquid-type startup rheostat; however such a method is expensive and complicated.

Finally, a synchronous motor can be turned over by an auxiliary motor with lower horsepower. In this case the startup must be done without load.

Synchronized Asynchronous Motors

These motors are used rather rarely despite the advantages they offer with respect to high power factors. They are wound-rotor motors synchronized by means of a nonalternating continuous excitation current to the rotor. Such motors have disadvantages. Sized with the larger air gap typical of synchronous motors, they can call for a large amount of current from the circuit in the event of a load loss, thus leading to a temporary disconnection. Also, in order to avoid excessive voltage at the rotor during startup the number of rotor coils is reduced, leading to a low level of field excitation and an awkward, expensive motor.

6   Drivers for Rotating Equipment

Efficiencies of Synchronous Motors

Typical efficiencies of synchronous motors are shown in Table 1-3. These efficiencies are reduced 0.5%-2% for low-speed motors. We have not distinguished between high voltage and low voltage, because these efficiencies are the same for units over 100 kW; below this, the high voltage motors have an efficiency reduced by 1%-3%.

Nonsynchronous vs. Synchronous Motors

Because of the small air gap in nonsynchronous motors, they are more fragile than synchronous motors in large sizes. And because of their superior power factors, synchronous motors are preferred over nonsynchronous motors for low speeds.

In dusty atmospheres nonsynchronous motors with squirrel-cage rotors stand up better than motors with rings.

Contemporary synchronous motors designed for dangerous atmospheres no longer show rings and brushes. The field excitation is achieved with an auxiliary alternator whose excitation is produced by a fixed coil; and the sinusoidal voltage generated by the winding of the alternator's rotor is then rectified by a bridge of turning diodes.

VARIABLE-SPEED ALTERNATING CURRENT MOTORS

Nonsynchronous Motors

The rotating speed, N, of a nonsynchronous motor with the number of poles, p, operating on a current with an alternating frequency, f, (60 or 50 cycles/sec) is given by the formula:

$$N = 160 f (1 - g)/p$$

Table 1-3
Typical Efficiencies for Synchronous Motors

Power (kW)	4/4 Load	3/4 Load	1/2 Load
25-50	90	88.5	85
75-100	92	91	88
150-200	93	92	89.5
More than 200	94.5	93.5	92

where g = slippage of the conductor through the magnetic field

Accordingly, there are two ways of varying the speed of a nonsynchronous motor:

1. By feeding power at a variable frequency.
2. By promoting slippage as the stator's frequency is kept constant.

Slippage in nonsynchronous motors can be varied by

1. Varying the stator's voltage for squirrel-cage motors.
2. Varying the current in wound rotors.

The latter of these two methods has been done for many years by means of a rheostat that changes the rotor resistance, although the resulting increased slippage is reflected in increased losses in the rotor and a limited depth of regulation. Consequently the contemporary trend is toward frequency variation.

Also, it is economical to regulate the speeds of large (several megawatts) induction motors with rings by means of a hyposynchronous cascade that recovers the energy of slippage dissipated in the rotor.

The Hyposynchronous Cascade

The circuitry for a hyposynchronous cascade is shown in Figure 1-2. This system controls the speed, N, of the rotor by supplying it enough current to produce a torque equal to the resisting torque at the desired speed.

In Figure 1-2, the rotor's voltage is continuous and adjustable through a diode bridge (3) fed by a thyristor bridge (4), which operates as an inverter and is connected to the fixed-frequency alternating current supply circuit. A transformer (5) adjusts the voltage of the supply circuit to that of the inverter. In order for the rotor to slip, the voltage from the inverter is varied by means of the delay angles of the thyristor bridge. An increase in voltage reduces the rectified rotor current and consequently the torque, thus causing the rotor to slow down until it reaches an equilibrium speed fixed by the demand torque, and vice versa.

The delay angle of the switches of the inverter thyristor bridge is determined by an output signal of an intensity regulator, which measures the rotor current at any moment. This regulator is in turn activated by an output signal of the speed regulator, which is set to regulate to the speed desired. This is thus a cascade controller.

This combination of diode, thyristor inverter, and transformer (3, 4, and 5 in Figure 1-2) returns to the rectifiers and transformer the slippage

8　Drivers for Rotating Equipment

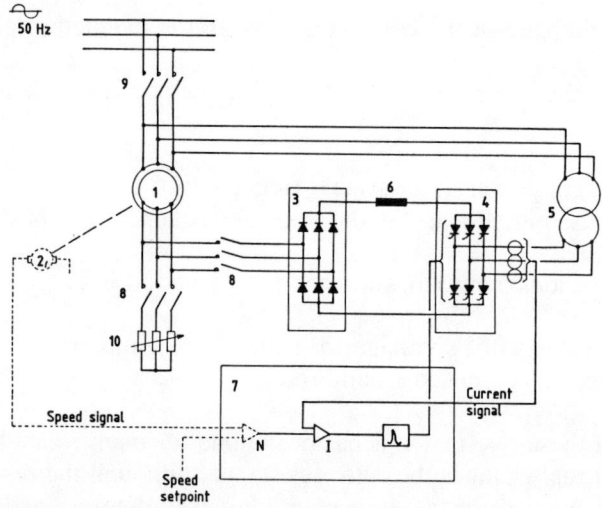

1 - Squirrel cage motor with rotor rings
2 - Tachometric dynamo
3 - Diode rectifier bridge ⎫
4 - Thyristor bridge   ⎬ energy recovery system
5 - Transformer adaptater   ⎪
6 - Smoothing inductance   ⎭
7 - Thyristor speed and initiation controller
8 - Rotor starting and rectifier - inverter coupling switches
9 - General protection equipment
10- Starting rheostat

Device groups (3) to (10) are assembled in electric panel or rack

The dashed lines represent   optimal equipment :
  tachometric dynamo - speed control

N = Instantaneous electric motor speed
Ns= Synchronous speed

$$S = \frac{Ns - N}{Ns} = \text{slipping}$$

I = Current intensity in rotor on speed N

Figure 1-2. Hyposynchronous cascade. (Courtesy of Jeumont-Schneider.)

energy that would be dissipated into heat with the traditional rheostat. The resulting efficiency (ratio between mechanical power at the shaft to the power taken from the network) is high, on the order of 0.90–0.94, depending on the size of the unit.

Speed Regulation with Variable Frequencies

Variable frequency is supplied to nonsynchronous motors by means of electronic frequency converters that additionally regulate the amplitude of the voltage in order to achieve constant flux. Numerous converters have been experimented with. The majority achieve an indirect conversion of frequency with a continuous-current intermediate stage. The only system capable of making a direct conversion of the alternating-current frequency is the "cycloconverter" (Figure 1-3). In this cycloconverter, each phase of the motor is fed from the three-phased supply circuit across two three-phased Graetz bridges mounted front-to-back in a group. The groups are each connected as part of a star to a separate winding of the motor. The neutral of the Graetz bridges is tied to the neutral of the stars (Figure 1-3). And the free end of each group is tied to a free end of a motor phase.

The average rectified voltage delivered by a Graetz bridge is a function of the delay angle of the switch to the thyratron of the bridge. For a given phase, therefore, the current alternates between the front bridge and the back bridge according to the variation of the delay angle, so that the average voltage delivered by the group has the shape of a definite sinusoid of

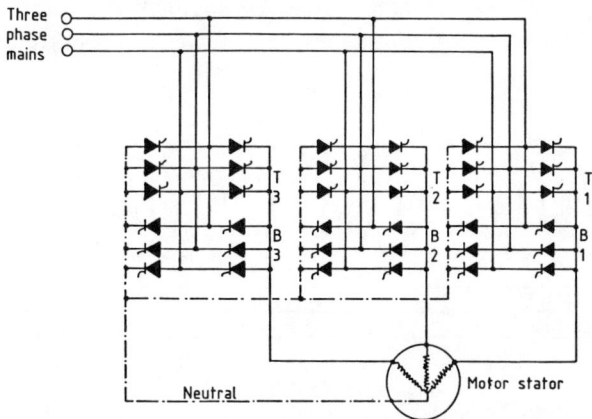

Figure 1-3. Circuitry for the cycloconverter speed regulator. (Courtesy of Jeumont-Schneider.)

10   Drivers for Rotating Equipment

determined amplitude and frequency (Figure 1-4). This represents the voltage obtained at the terminal of one phase of the motor. In this example, the amplitude of the sinusoidal voltage curve is the maximum and equal to the voltage amplitude of the feed circuit.

The current in any one phase has a form close to a sinusoid since it is dampened by the inductances of the motor. The result is a three-phased system suitable to feed the motor by controlling the three Graetz-bridge groups according to the same fixed initiation, but with a time lag equal to one third of the period of the desired outlet frequency. Figure 1-4 shows that the outlet voltage is obtained by sampling the inlet voltage and it appears that this sampling gets better when the outlet frequency is smaller.

Practically, the already described cycloconverter with natural commutation, which is the one most used, limits excursions of the outlet frequency to between zero and one-third the inlet frequency. This frequency conversion is a good solution for large motors of several MW power, operating at slow speeds of 30–500 rpm. Then the efficiency is excellent, on the order of 97%.

However, the present most commonly used speed regulator is the rectified voltage commutator or rectified continuous-voltage inverter operating on alternating voltage input. This is a static frequency converter with an intermediate-ring continuous-current circuit, which includes four essential segments (Figure 1-5):

1. A Graetz bridge connected to the circuit with 6 thyratrons and operating as a pilot rectifier whose voltage in amplitude is regulated according to the delay in charging the thyratrons (Figure 1-5). A delay of 0° gives maximum average voltage, and a delay of 90° gives negligible average voltage. This first segment regulates the amplitude of the voltage to the motor as a function of the frequency and

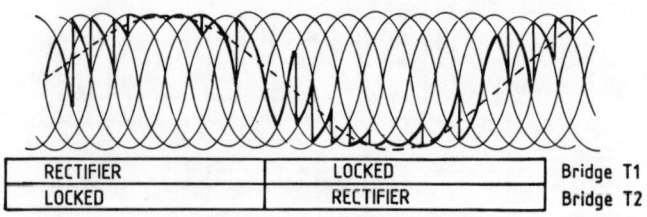

TERMINALS VOLTAGE OF ONE PHASE

Figure 1-4. The sinusoidal curve of rectified voltage from a Graetz bridge. (Courtesy of Jeumont-Schneider.)

Electric Motors   11

thus of the desired speed, in such a way as to maintain the flux of voltage-to-frequency constant.

2. A filtering cell located in the continuous current circuit (Figure 1-5). This filtering cell comprises an inductance, L, and capacitance, C, and functions to dampen the "continuous" voltage delivered by the rectifier, which in fact oscillates at 300 Hz.

3. A three-phase Graetz bridge with six thyratrons, operating as a variable-frequency autonomous inverter that feeds the asynchronous motor with a three-phased system of alternating voltages in shifted steps each in relation to the other for a third of the period. In order to create these voltages from the continuous voltage delivered by the first segment (Figure 1-5), the inverter acts as a cyclic commutator to alternately apply the plus-minus polarities on each terminal of the motor. It does this by periodically charging and extinguishing the bridge's six thyratrons successively, according to a predetermined order. The frequency of the voltage to the motor is determined by the frequencies of the thyratron switches.

4. A device for switching off the thyratrons. This is indispensable for the commutation in the case of an asynchronous motor, or of a synchronous motor operating with lagging phase angle. Because the inverter, segment 3, which is the source of alternating current for the nonsynchronous motor, acts as a passive charge with no alternating voltage of its own, this inverter is capable of provoking a natural commutation of the thyratrons; and it is necessary to force the commutation by forcing the switch-off of the thyratrons of the inverter bridge. This is the source of the name "inverter with forced commutations" or "auto-commuted inverter" which is given this bridge. There are several ways of having a switch-off device; Figure 1-5 gives one example based on auxiliary thyratrons. If the sys-

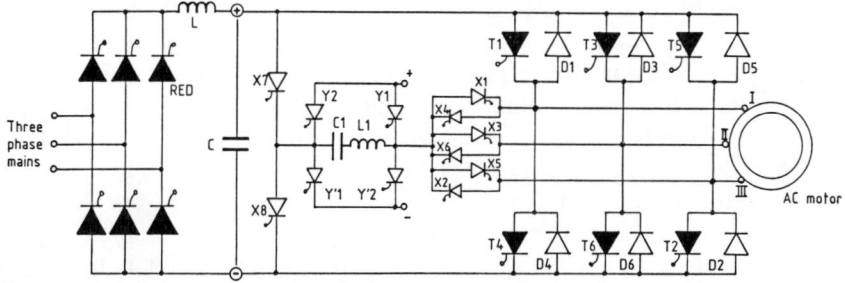

Figure 1-5. Rectified voltage commutator speed-regulator. (Courtesy of Jeumont-Schneider.)

12   Drivers for Rotating Equipment

tem is controlled by a speed regulator, the voltage control is based on the desired speed with intensity limitation from the pilot rectifier and the frequency control from the inverter. A correction can be supplied at low speeds holding to a constant flux ratio of voltage to frequency.

The overall efficiency of this system is good, ranging from 80% at 20 Hz to 91% at 50 Hz. Also, this system affords starting the motor at nominal torque without over-voltage, by feeding the stator with reduced frequency (about 1 Hz) and with the voltage correspondingly reduced to give a constant voltage-to-frequency flux ratio.

This system is well suited to services where continuous-current motors cannot be used for reasons of high speed, of constant maintenance, or of hazardous environment (for example, the presence of a mechanical commutator is excluded in explosive atmospheres). However, the expense of the forced commutation device penalizes this system, which is rarely used with low voltages above 500 kW. For economic reasons the autopiloted synchronous motor is often preferred.

Synchronous Motors

The synchronous motor offers two outstanding characteristics:

1. Because it does not experience slippage, it affords rigorous speeds from electronic clocks (quartz oscillators).
2. Because it can act on the circuit as a capacitor, it can make use of converters with natural commutation.

The machines used for variable speed are generally motors without commutator rings or brushes, the rotor winding being fed (for high power machines of several MW) by continuous current from a rectifier with diodes, turning with the rotor in conjunction with a turning nonsynchronous generator or transformer. Regulation of the field current is done with a thyratron graduator placed between the three-phased supply circuit and the fixed part of the field exciter.

There are two very different ways of variable-speed control of synchronous motors:

1. A source with variable frequency and voltage as for nonsynchronous motors.
2. Auto-piloting the current supply by the motor, which then acts as a continuous current motor or an auto-synchronous motor.

The first of these two holds little interest, and we will examine the second only.

The system consists of, for example, two static converters in a three-phased Graetz bridge connected through an intermediate stage with continuous current dampened by an inductance (Figure 1-6). One of the converters is connected to the circuit and operates as a rectifier (Figure 1-6) to deliver continuous current to the intermediate circuit. The other converter is connected to the motor and operates as an inverter (Figure 1-6); it periodically commutes the continuous current from one phase of the synchronous motor to another, thus supplying its stator with a three-phased alternating current whose variable frequency is defined by the frequency of commutation of the thyratrons of the inverter.

In synchronous mode, as soon as the synchronous motor turns at sufficient speed, the inverter's commutations are caused by the voltage at the motor's terminals, so that the motor furnishes the necessary reactive energy. These commutations are controlled by a position sensor, called an angle coder, which comprises a slotted disc attached to the rotor that passes before sensors fixed on the stator, so as to start each thyratron the instant the polar axis takes a prescribed position relative to the coils of the stator.

The commutations of the inverter are thus given their rhythm by the rotations of the motor itself, and the supply frequency of the motor consequently remains always perfectly synchronous with the motor's speed. This is why the synchronous motor is said to be auto-piloted or autosynchronous.

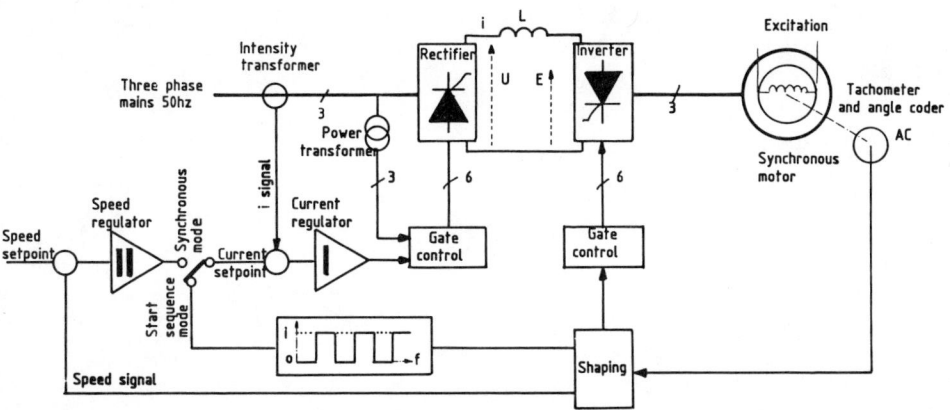

Figure 1-6. Synchronous motor self piloted by an angle coder. (Courtesy of Jeumont-Schneider.)

14 Drivers for Rotating Equipment

This method avoids any risk of disconnecting the machine. Any slowdown of the motor, whether abrupt or gradual, automatically leads to a corresponding decrease in the frequency of the supply currents.

In sequential mode, at shutdown or low speed, the motor is not capable of causing the commutation of the inverter; and natural commutation is possible only when the motor's speed is above a minimum on the order of 8% of the nominal, corresponding to the short-circuit voltage. Below this speed, the thyratrons of the inverter (Figure 1-6) are operated in forced commutation by the angle coder, which activates the previously conducting thyratrons by cutting off the continuous current in the intermediate circuit, and then charging the new thyratrons before connection. The passage to synchronous mode is automatic when the speed gets high enough.

In order to avoid an unacceptable over-voltage, the rectifier (Figure 1-6) should deliver a voltage well adjusted to the motor's reaction voltage, which is proportional to its speed. The rectified voltage must therefore vary proportionally to the frequency and the rectifier bridge is con-

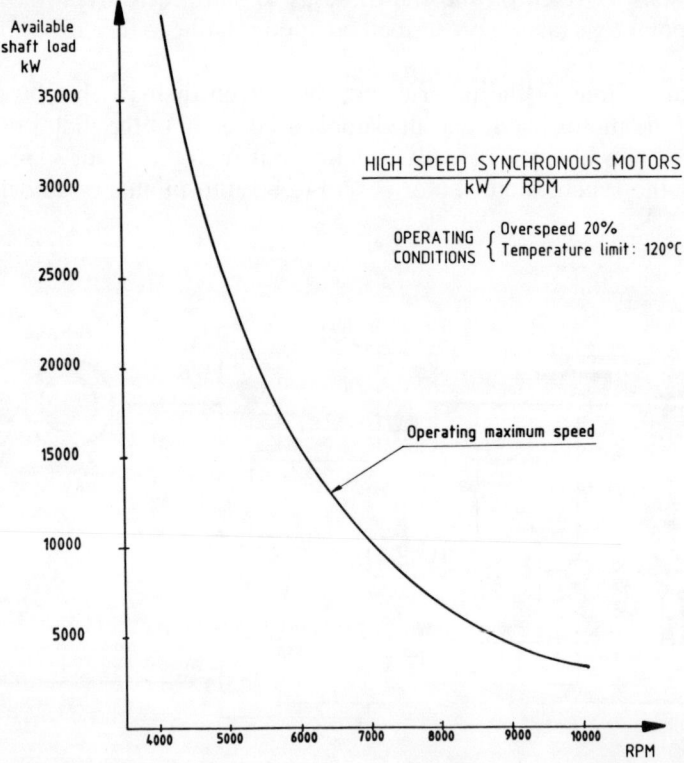

Figure 1-7. Variation of maximum speed with rated power of synchronous motors.

sequently regulated by the switching delay of its thyratrons piloted by the frequency of the supply circuit. This regulation is of the cascade type; the speed regulator on which the desired command is posted, furnishes the current reference to the intensity regulator piloting the gate controls of the rectifier bridge.

The power of these motors, which are used for driving compressors, can vary from 2 to 30 megawatts. The combined efficiency of the transformer + converter + motor is a function of the nominal power and ranges from 95% for 25 MW motors to 91% for 2 MW motors running at nominal speeds. Such combined efficiencies are a product of the separate efficiencies of the parts, which are: 99% for the transformer, 97%-98% for the converter, and 94%-96% for the motor.

The maximum speeds achieved with these motors are 6,500 rpm for 15 MW motors to 8,000 rpm for 4 MW motors. Figure 1-7 shows the variation of maximum speed with rated power.

It must be noted that auto-piloted synchronous motors without rings or brushes are limited to speeds of 6,500 rpm for maintaining the field-excitation diodes. For higher speeds, the commutator rings and brushes are retained, which however, requires more maintenance.

The cost of such variable-speed equipment is twice the cost of a constant-speed motor for powers of 5-10 MW; and the multiplying coefficient goes from 2 to 2.5 and even 3 as the power is reduced below 2 MW.

Chapter 2
General Characteristics of Heat Engines

Heat engines convert the energy of fired heat to mechanical energy through combustion in confined spaces. We here distinguish five kinds of such machine:

1. Spark-ignited internal combustion engines
2. Auto-ignited internal combustion engines
3. Mixed-cycle internal combustion engines
4. Gas turbines
5. Steam turbines associated with steam generators and condensers

In spark-ignited engines, the fuel and air have been mixed in a sealed space at the moment of ignition, which is provoked by an electric spark. The consequent very rapid combustion occurs at theoretically constant volume (Beau de Rochas cycle). In auto-ignition engines, the fuel is injected into an air mass already heated by compression in a sealed space; the fuel flashes spontaneously; and the combustion occurs at theoretically constant pressure (Diesel cycle). In mixed-cycle engines, the combustion in the sealed space occurs first at constant volume, then at constant pressure (Sabathe cycle). In reality this mixed cycle occurs in all engines, whether ignition is controlled by a spark or by auto-ignition; constant-volume combustion is merely preponderant with controlled ignition, and constant-pressure merely is preponderant when the injection of fuel before the high-compression dead-point initiates the combustion. In all cases, however, the combustion gases expand in a positive displacement chamber.

By contrast to the closed cylinders of the foregoing engines, the fuel for the gas turbines is injected continuously into a combustion chamber located at the discharge of an air compressor; the hot gases expand into a turbine driving both the air compressor and the connected load. The combustion occurs and the gases expand at a constant pressure equivalent to the inlet pressure of the air compressor.

THERMAL EFFICIENCY OF A HEAT ENGINE—SPECIFIC CONSUMPTION

The overall thermal efficiency of a heat engine, Rg, is the ratio of the energy released at the shaft at constant conditions during a number of complete cycles to the mechanical equivalent of the heat energy liberated by the fuel during a complete oxidation at constant volume without credit for the heat of condensation of water in the combustion products (LHV). This can be expressed mathematically as:

$$R_g = \frac{\Im e}{J \, m \, LHV}$$

where $\Im e$ = mechanical energy recovered at the shaft, kj
 LHV = low heating value of the fuel, kcal/kg
 R_g = overall efficiency
 m = mass of fuel, kg
 J = mechanical equivalent of heat, 4.184 kj/kcal

Also,

$$R_g = \eta \cdot \eta_s \cdot \eta_{co} \cdot \eta_m$$

where η = efficiency of the ideal theoretical cycle
 η_s = efficiency of the actual cycle relative to that of the theoretical cycle
 η_{co} = efficiency of the combustion reaction
 η_m = mechanical efficiency of the engine

Thus the specific consumption of a heat engine is used to characterize the overall thermal efficiency, as:

- kcal/kWh or kJ/kWh
- Btu/BHPH for brake horsepower on the shaft
- g/kWh or g/hp for a typical fuel taken as having a heating value (LHV) of 10,000 kcal/kg

The effective work produced at the shaft is understood as the work produced by the engine proper minus that work consumed for driving auxiliaries such as cooling water pumps, oil pumps, air superchargers, and so forth.

18 Drivers for Rotating Equipment

Typical overall thermal efficiencies are as follows:

- Automobile engine: 25%–27%
- Large supercharged gas engine of 1,500–9,000 kW: 36%–38%
- First generation gas turbine: 22%–27%
- Second generation gas turbine: 30%–36%
- Steam cycle involving a fired boiler, turbine, and vacuum condenser: 22%–35%
- Steam cycle involving a fired boiler, back-pressure turbine, and use for heat of exhaust steam: 40%–90%

THEORETICAL THERMODYNAMIC EFFICIENCY

Assume 1 kg of fluid passing through a complete cycle (Figure 2-1):

$$\eta = \frac{\int_c p \cdot dv}{JQ}$$

where Q = the quantity of heat released by the combustion during the cycle

The cycle shown in Figure 2-1 can be differentiated into an infinite number of differential Carnot cycles, each with an efficiency, $\eta = 1 - (T_1/T_2)$, closer to 1 as the ratio of T_1/T_2 approaches zero. Consequently, the efficiency of a heat engine of any type tends to be higher as the combustion temperature is higher and as the temperature of the exhaust is lower. Similarly, the efficiency of a condensing-turbine steam cycle gets higher as the steam temperature gets higher and as the condenser temperature gets lower (i.e., vacuum condensers).

Also, the theoretical thermodynamic efficiency improves as the cycle approaches a perfect Carnot cycle of two adiabatic and two isothermal transitions, at which the Carnot cycle exhibits its maximum efficiency between T_1 and T_2.

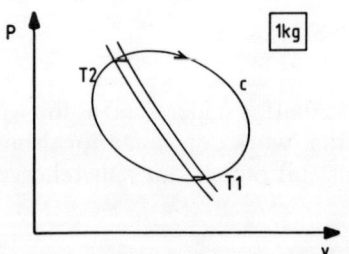

Figure 2-1. General cycle.

Theoretical Thermodynamic Efficiency of a Constant-Volume Heat Engine (Beau de Rochas Cycle, Figure 2-2)

$$\eta = \frac{Q_1 - Q_2}{Q_1}$$

where Q_1 = heat released from Point 2 to Point 3 in Figure 2-2
Q_2 = heat lost in the exhaust, as represented between Points 4 and 1 in Figure 2-2.

The energy delivered by the engine is $Q_1 - Q_2$. Also, the efficiency and temperatures are related through the constant-volume expansions as follows:

$$\eta = 1 - \frac{Q_2}{Q_1}$$

$$\eta = 1 - \frac{c_v (T_4 - T_1)}{c_v (T_3 - T_2)}$$

Where c_v is the average specific heat of the air-fuel mixture before and after combustion at constant volume. Then:

$$\frac{T_2}{T_1} = \left(\frac{p_2}{p_1}\right)^{\frac{\gamma-1}{\gamma}} = \left(\frac{v_1}{v_2}\right)^{\gamma-1}$$

$$\frac{T_3}{T_4} = \left(\frac{p_2}{p_4}\right)^{\frac{\gamma-1}{\gamma}} = \left(\frac{v_4}{v_3}\right)^{\gamma-1}$$

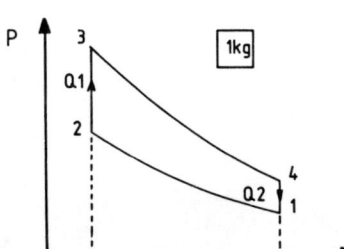

Figure 2-2. Beau de Rochas cycle.

but . . .

$$\frac{V_1}{V_2} = \frac{V_4}{V_3} = \frac{V}{v} = \rho$$

where ρ is the volumetric ratio, and often assumed to be the compression ratio for gasoline engines. Therefore:

$$\eta = 1 - \frac{T_4 - T_1}{\rho^{\gamma-1}(T_4 - T_1)}$$

$$\eta = 1 - \frac{1}{\rho^{\gamma-1}}$$

Experience has shown that $\gamma = 1.27$ for gasoline engines. The volumetric ratio is usually wrongly called the compression ratio.

Theoretical Thermodynamic Efficiency of a Constant-Pressure Heat Engine (Diesel Cycle)

$$\eta = 1 - \frac{Q_2}{Q_1}$$

$$\eta = 1 - \frac{c_v(T_4 - T_1)}{c_p(T_3 - T_2)}$$

where c_p = average specific heat of the fuel and air mixture before and after combustion at constant pressure
Q_1 = heat released from Point 2 to Point 3 in Figure 2-3.

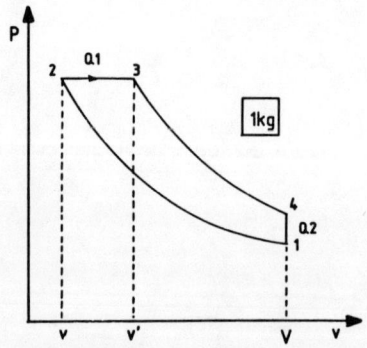

Figure 2-3. Diesel cycle.

$$p_1 v_1 = r T_1$$

$$p_2 v_2 = r T_2$$

$$p_3 v_3 = r T_3$$

$$p_4 v_4 = r T_4$$

$$\gamma = \frac{c_p}{c_v}$$

$$\eta = 1 - \frac{1}{\gamma} \frac{p_4 v_4 - p_1 v_1}{p_3 v_3 - p_2 v_2}$$

$$\eta = 1 - \frac{1}{\gamma} \frac{v_1 (p_4 - p_1)}{p_2 (v_3 - v_2)}$$

Referring to Figure 2-3, let

$$\frac{V}{v} = \rho$$

Then let

$$\frac{V}{v'} = \rho'$$

$$\frac{v'}{v} = \frac{\rho}{\rho'} = \epsilon$$

where ϵ is the expansion during combustion. From this:

$$\eta = 1 - \frac{1}{\gamma} \frac{\dfrac{p_4}{p_2} - \dfrac{p_1}{p_2}}{\dfrac{v_3}{v_1} - \dfrac{v_2}{v_1}}$$

$$\eta = 1 - \frac{1}{\gamma} \frac{\left(\dfrac{1}{\rho'}\right)^{\gamma} - \left(\dfrac{1}{\rho}\right)^{\gamma}}{\dfrac{1}{\rho'} - \dfrac{1}{\rho}}$$

$$\eta = 1 - \frac{1}{\gamma} \frac{1}{\rho^{\gamma-1}} \frac{\epsilon^\gamma - 1}{\epsilon - 1}$$

Experience has shown that $\gamma = 1.35$ for diesel oil; ϵ is close to 1 at low loads; and $\epsilon > 1$ as soon as there is a load.

Efficiencies of the Beau de Rochas and Diesel Cycles

For engines with spark ignition, the volumetric ratio, ρ, is limited to 9–10, because higher volumetric ratios lead to auto-ignition and knocking. For a Diesel engine, the volumetric ratio ρ must be more than 12 in order to achieve auto-ignition and can reach 25. Accordingly, typical volumetric ratios and thermodynamic efficiencies are as follows:

- Beau de Rochas cycle:

$\rho = 8$

$\eta = 44\%$

- Diesel cycle:

$\rho = 20$

$\eta = 65\%$

Thermodynamic Efficiency of a Mixed-Cycle Engine (Actual Cycle for Spark Ignition or Diesel)

The cycle diagram is divided into a diagram of the Beau de Rochas and the diesel cycles (Figure 2-4). Relating terms from this diagram:

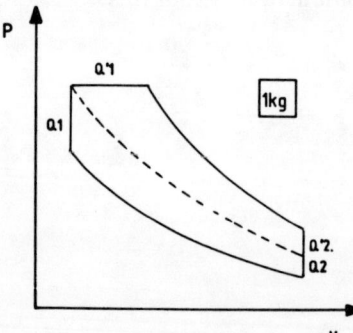

Figure 2-4. Combined cycle.

$$\eta_1 = \frac{Q_1 - Q_2}{Q_1}$$

$$\eta_2 = \frac{Q'_1 - Q'_2}{Q'_1}$$

$$\eta = \frac{Q_1 - Q_2 + Q'_1 - Q'_2}{Q_1 + Q'_1}$$

$$\eta = \frac{\eta_1 Q_1 + \eta_2 Q'_1}{Q_1 + Q'_1}$$

Thermodynamic Efficiency of a Gas Turbine

The thermodynamic efficiency of a gas turbine can be determined from assuming the cycle shown in Figure 2-5, as follows:

$$\eta = \frac{Q_1 - Q_2}{Q_1}$$

$$\eta = 1 - \frac{c_p (T_4 - T_1)}{c_p (T_3 - T_2)}$$

$$\gamma = \frac{c_p}{c_v}$$

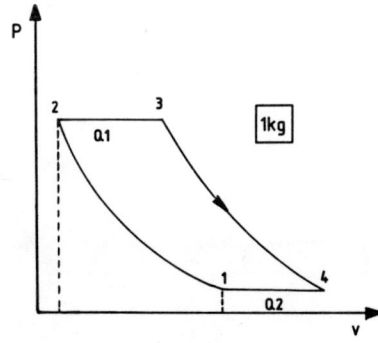

Figure 2-5. Gas turbine cycle.

24 Drivers for Rotating Equipment

$$\frac{T_2}{T_1} = \left(\frac{p_2}{p_1}\right)^{\frac{\gamma-1}{\gamma}} = r^{\frac{\gamma-1}{\gamma}}$$

$$\frac{T_3}{T_4} = r^{\frac{\gamma-1}{\gamma}}$$

$$\eta = 1 - \frac{1}{r^{\frac{\gamma-1}{\gamma}}}$$

If the maximum combustion temperature, T_3, is set by the materials of the combustion chamber, turbine vanes, blades, and so forth, plus the temperature of the ambient air, T_1, the maximum recoverable work can be derived as follows:

$$r = \left(\frac{T_3}{T_1}\right)^{\frac{1}{2}\frac{\gamma-1}{\gamma}} = \left(\frac{T_2}{T_1}\right)^{\frac{1}{\frac{\gamma-1}{\gamma}}}$$

To simplify:

$$\frac{\gamma-1}{\gamma} = \alpha$$

The expression for work is:

$$Q_1 - Q_2 = c_p (T_3 - T_2) - c_p (T_4 - T_1)$$

$$= c_p \left(T_3 - T_1 r^\alpha - \frac{T_3}{r^\alpha} + T_1\right)$$

And the maximum work is when:

$$\frac{d(Q_1 - Q_2)}{dr} = 0$$

$$-\alpha T_1 r^{\alpha-1} + \alpha T_3 r^{-\alpha-1} = 0$$

$$r = \left(\frac{T_3}{T_1}\right)^{\frac{1}{2\alpha}}$$

General Characteristics of Heat Engines

This result is also true for reciprocating engines. It can be seen that an increase in T_3/T_1 translates to an increase in compression ratio, r, and therefore to an increase of efficiency as indicated earlier.

In conclusion, the efficiency of a heat engine increases with the volumetric ratio, ρ, therefore with the compression ratio, r, and consequently with the ratio of the combustion temperature before expansion to the temperature of the inlet air.

SHAPE EFFICIENCIES FOR THERMODYNAMIC CYCLES (η_s)

Assume that the shape efficiency of a real cycle, η_s, is equal to the ratio of the efficiency of the real cycle, η', to that of the corresponding theoretical cycle, η, or $\eta_s = \eta'/\eta$.

In order to determine η_s, it is necessary to obtain the actual cycle diagram, and that diagram's area is measured, because:

$$\eta' = \frac{\int_{c'} p \cdot dv}{Q_1}$$

In the case of gas turbines, it is easy to determine the thermodynamic efficiency corresponding to the actual cycle, η', as the actual cycle is distorted by losses in the compressor and turbine. In the case of gasoline engines and diesel engines, the actual diagram is distorted by the following seven phenomena:

1. Combustion does not occur at the theoretical point, and the ignition is regulated ahead or delayed to have the best possible combustion for the speed, the load, and the fuel.
2. The combustion is difficult to control, especially in spark ignited engines.
3. A secondary combustion often occurs after the end of the theoretical combustion, depending on the composition of the fuel/air mixture.
4. Variations occur in the pressure drops across the valves or ports as well as at the inlet and exhaust.
5. The speed of combustion can vary as a function of the shape of the combustion chamber and the spark-plug location.
6. Variations occur in the fuel/air mixture and its richness (fuel-air equivalence ratio).
7. Variations occur in the fuel atomization.

The fuel-air equivalence ratio or richness is the ratio of fuel/air over fuel/air in stoichiometric conditions. The inverse is the air-fuel equivalence ratio, also designated λ.

COMBUSTION EFFICIENCY

The combustion in positive displacement engines is never perfect; and the theoretical heat of combustion is not reflected in either the composition or the temperature of the combustion gases. The combustion efficiency, which is the ratio of the actual to theoretical heat release, is affected by five things:

1. The ratio of surface to volume in the combustion cylinder. Small motors have a poorer efficiency than do large ones.
2. The amount of cylinder wall cooling. Losses through the walls get larger as the temperature and pressure get higher, as the gas turbulence increases, and as the cylinders get smaller.
3. The fuel composition and richness. The richness should normally equal 1.0 in order to have the highest possible temperature, but the optimum richness also depends on the phenomena of detonation and the tendencies for pollutant in the exhaust.
4. The fuel atomization and dispersion. This should be the best possible for a homogeneous mixture of fuel + air.
5. The turbulence. High turbulence favors homogeneous combustion.

In spark-ignited engines liquid fuels are sprayed into the cylinder by a carburetor or an injection system, and are subsequently vaporized before combustion by the heat remaining in the cylinder. The gross quantity of heat available after the combustion is therefore higher than the LHV by the latent heat of vaporization of the fuel, λ, and the efficiency is multiplied by $(LHV + \lambda)/LHV$. For diesel and gas engines there is naturally no increase in the thermal efficiency reported to LHV.

In the case of open-cycle gas turbines the combustion is carried out with a large excess of air, and can be considered as complete. As with cylinder engines, good fuel atomization is necessary, additionally in the case of turbines, to avoid having droplets impinge on the blades and lead to post combustion that might destroy the blades. The heat of combustion serves both to vaporize the fuel and to raise the exhaust-gas temperature to the outlet temperature of the combustion chamber. This corresponds to a combustion efficiency for the increase in air temperature of around 0.95.

For gas fuel the combustion efficiency depends on the ratio of the upper explosion limit to the low explosion limit which must be greater than 3 to have an efficiency over 0.95 (Figure 2-6).

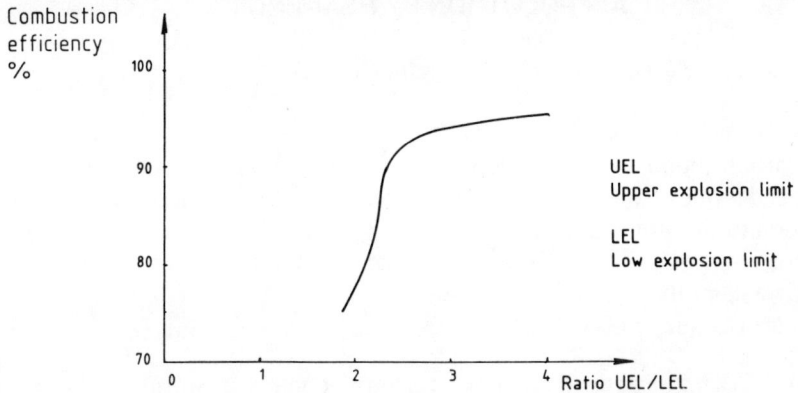

Figure 2-6. Combustion efficiency for gaseous fluids.

MECHANICAL EFFICIENCY (η_m)

Mechanical efficiency is the ratio of the actual energy conversion to the theoretical conversion for an ideal frictionless engine. The mechanical losses are transferred into heat that is absorbed by the lubricating oil, and cooling is an important function of the oil in addition to its regular function of lubrication. Mechanical losses are proportional to speed.

INDICATED EFFICIENCY (η_i)

A combined efficiency term used for reciprocating internal combustion engines is the "indicated" efficiency, η_i. The indicated efficiency, which can be read directly from an operating engine, equals the product of the thermodynamic, shape, and combustion efficiencies, or:

$$\eta_i = \eta \, \eta_s \, \eta_{co}$$

and the overall thermal efficiency is $R_g = \eta_i \, \eta_m$.

APPROXIMATE EFFICIENCIES FOR RECIPROCATING INTERNAL COMBUSTION ENGINES

$\eta_s = 80\% - 90\%$

$\eta_{co} = 60\% - 90\%$

$\eta_m = 85\% - 90\%$

AIR POLLUTION BY HEAT ENGINES

The exhaust gases from heat engines can contain:

- Unburned hydrocarbons (UHC).
- Carbon monoxide from incomplete combustion to carbon dioxide.
- Lead oxides from tetraethyl lead added to the fuel to increase resistance to auto-ignition (knocking).
- Nitrogen oxides (NO, NO_2, NO_x) due to combustion of nitrogen in the air and in the fuel.
- Sulfur oxides SO_x (SO_2, SO_3).

If the exhaust gases have no or very low content of sulfur oxides, then the fuel must have as little sulfur as possible to prevent corrosion. The content of SO_x can be determined by the formula SO_x (in g/kWh) $= 0.5$ weight % of S in fuel.

Sulfur oxides emissions are a one-to-one function of the sulfur input in the fuel and 3% by weight of the sulfur is converted into SO_3. When pipeline natural gas is burned, zero sulfur oxides are emitted.

The permissible levels of polluting agents in the exhaust get lower and lower, as anti-pollution laws get more severe (the methane CH_4 is not considered a pollutant).

The unburned hydrocarbons (UHC) must be reduced as much as possible, which represents an energy loss. Turbulence and secondary air injection is used in gas turbines to keep the combustion chamber from having any places where the fuel-air mixture is too lean or too rich. The unburned hydrocarbons are measured as CH_4 equivalent.

For gas turbines the UHC emission increases in a manner of a hyperbolic curve as the firing temperature decreases; however between 50% and 100% load, the UHC emissions level does not change (Figure 2-7).

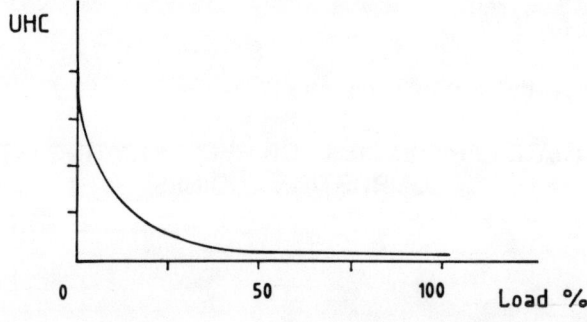

Figure 2-7. Unburned hydrocarbon emissions in a gas turbine.

Nitrogen oxides and the carbon monoxide are produced by the combustion which is adiabatic.

Carbon monoxide is produced if there is not enough oxygen during the combustion; the fuel-air equivalence ratio or richness is greater than 1.

The nitrogen oxides (NO_x) are the most toxic and are formed in the flame from

- The fuel bounded nitrogen (FBN) which is nearly equal to 10% of the sulfur in the complex fuel molecules; there is almost no FBN in gaseous fuels. It is the "organic NO_x."
- The nitrogen of the combustion air. It is the "thermal NO_x."

The organic NO_x is in small quantities and is formed in the flame zone where the combustion is intensive but the oxidation of fuel-bounded nitrogen has an efficiency of about 100% at low FBN content in fuel (till .05% by weight) and falls to 20% for a .6% content.

The thermal NO_x is formed as follows:

$$N_2 + O \rightarrow NO + N \qquad (2\text{-}1)$$

$$N + O_2 \rightarrow NO + O \qquad (2\text{-}2)$$

$$OH + N \rightarrow NO + H \qquad (2\text{-}3)$$

Reaction 2-3 is only important for mixtures of a fuel-air equivalence ratio markedly over 1.

These reactions were developed by Zeldovich and the NO formation velocity is an exponential function of the temperature.

The NO formation increases as

- The flame temperature increases. This temperature depends on the fuel-air equivalence ratio (Figure 2-8) and on the fuel type (Figure 2-9). NO increases exponentially with flame temperature.
- The combustion gases are maintained a long time at high temperature.
- The combustion pressure increases. NO content increases with the square root of the pressure.

The nitrogen oxide, which is formed in the flame (Figure 2-10) is transformed almost immediately by further oxidation into NO_2. The amount of NO_x in the exhaust is measured as dry NO_2.

The amount of NO_x formed depends on the fuel-air equivalence ratio or richness. It is the same for the CO and UHC formation (Figure 2-11).

30 Drivers for Rotating Equipment

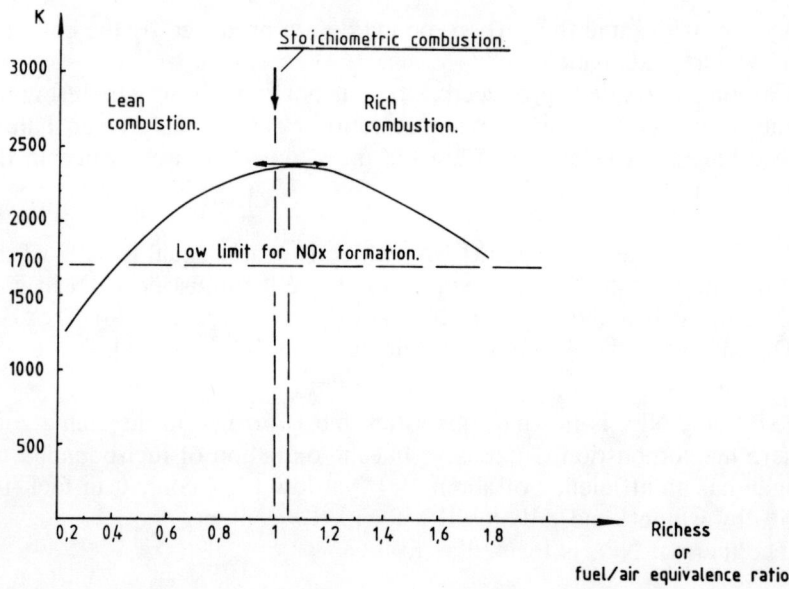

Figure 2-8. Flame temperature for air-methane mixture at 600 K (327°C).

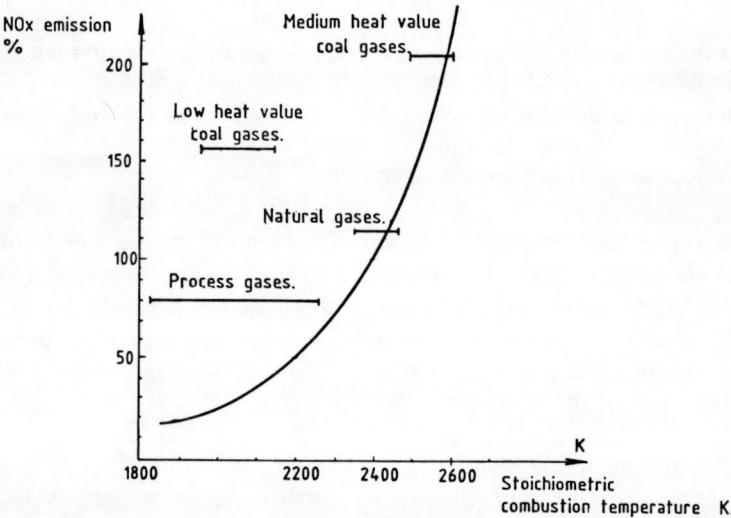

Figure 2-9. Influence of fuel and combustion temperature on NO_x formation in gas turbine exhaust.

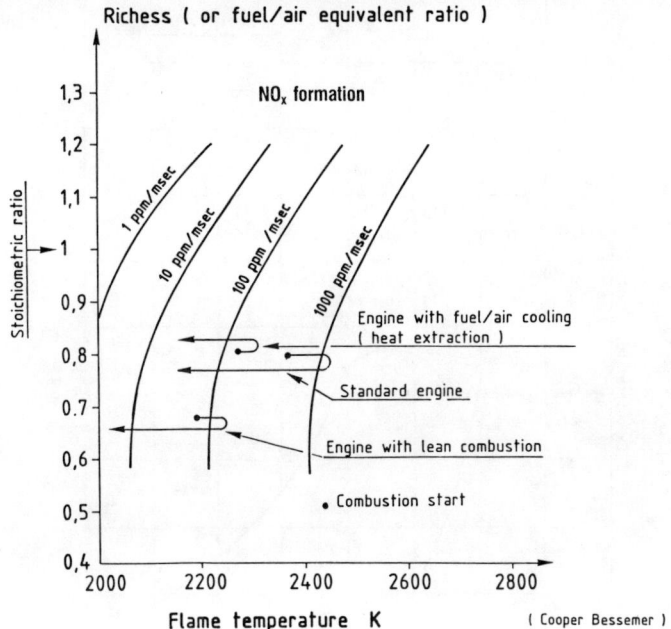

Figure 2-10. NO_x generation in heat engines. (Courtesy of P. R. Danyluk and F. S. Shaub, Cooper Industries.)

The polluting agents are measured as dry gases:

- In ppmv (parts per million in volume) or cm^3/m^3.
- In mg/nm^3 (m^3 at 0°C and 1,013 bar—760 mm Hg).
- In g/kWh.

The equivalences are

NO_x: 1 ppmv = 2.05 mg/nm^3

CO: 1 ppmv = 1.25 mg/nm^3

UHC: 1 ppmv = 0.71 mg/nm^3 (as CH_4)

The mg/kWh depend on the thermal efficiency of the engine and on the excess air, combustion air, and dilution air for gas turbines or scavenging air for two-cycle engines:

32 Drivers for Rotating Equipment

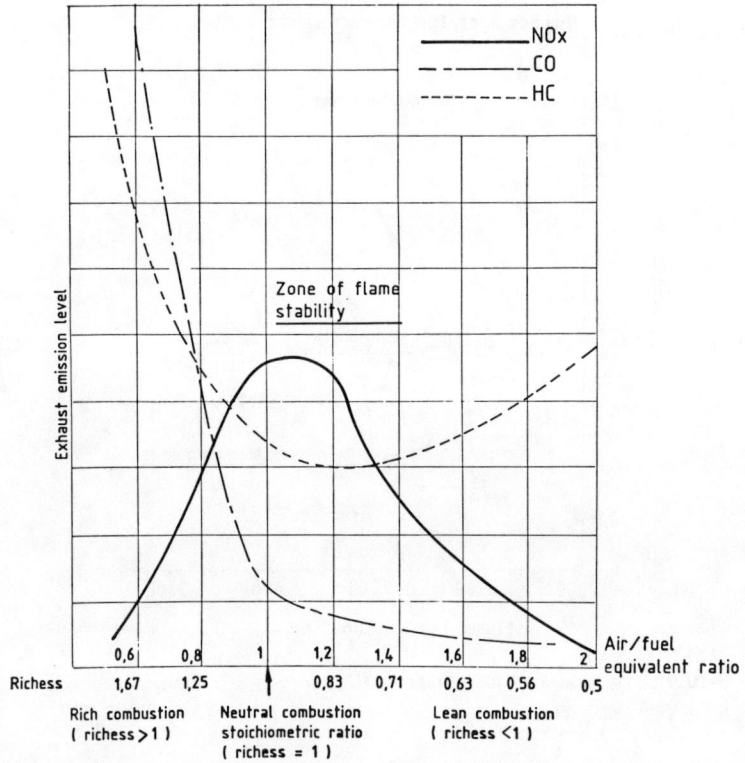

Figure 2-11. Exhaust pollutants as functions of fuel/air richness. (Courtesy of P. R. Danyluk and F. S. Shaub, Cooper Industries.)

$$\text{mg/kWh} = \text{mg/nm}^3 \cdot \frac{1}{\text{LHV}} \cdot \frac{1}{\eta_{th}} \left[V + v\left(\frac{1}{r} - 1\right) \right]$$

where LHV = low heating value of the fuel
r = richness (fuel-air equivalence ratio for the total air)
η_{th} = thermal efficiency related to LHV
v = air volume
V = exhaust gas volume } for a richness of 1 and 1 kg of fuel

In the United States, the NO_x content of gas turbine exhausts has been limited since 1973 for new turbines with a consumption of more than 50 million Btu/hr, i.e., more than 3,000–4,000 kW power, as follows: 225 ppm by volume of NO_2 for an oxygen content of 3 vol.%. This corresponds to 75 ppm NO_x with 15% oxygen, which is the conventional oxygen level for a gas turbine, instead of 3%, which is typical of a piston engine.

The NO_x content is related to the content at 3% oxygen from an analysis of the exhaust gas by assuming that the air is an inert diluent as follows:

$$(NO_x \text{ at } 3\%) = (NO_x \text{ at } x\%)(21 - 3)/(21 - x)$$

For gas turbines with a consumption of less than 50 million Btu/hr the limit is 150 ppmv with 15% oxygen

At the end of 1977, the U.S. Clean Air Act was modified and imposed an annual limit of 100 tons per year of polluting agents for a stationary engine or turbine, with a transitory limit of 250 tons per year on the understanding that manufacturers should study their engine designs to attain the 100-ton limit. This limit of 100 tons also applies to fixed installations such as steel mills and chemical plants.

In California the NO_x content is limited to 42 ppmv at 15% oxygen.

In some countries the CO and UHC contents are also limited.

Machines not suited to a reduction of NO_x in their emissions typically have the following NO_x exhaust levels:

- Gas turbines: 90–450 ppmv with 12% to 18% oxygen.
- Reciprocating engines: 1,000–1,200 ppmv with a very low oxygen content on the order of 2%–3%.

Gas turbines susceptible to emissions improvement actually discharge pollutants as shown in Table 2-1. Gas engines susceptible to emission improvement actually discharge pollutants as shown in Table 2-2. It can be seen that for both gas turbines and gas engines the most important pollutant is NO_x, which is 1–2.7 g/kWh for gas turbines and 6–35 g/kWh for gas engines.

The NO_x content of industrial gas engines is about double that of gasoline automobile engines, because of:

- Low cooling losses during combustion due to the large-volume cylinders and relatively smaller ratio of wall surface to volume.
- The higher temperature (550°C) of the air-fuel mixture at the moment of ignition due to supercharging.
- An air-fuel equivalent ratio of 1.15–1.28, which corresponds to the maximum formation of NO_x.
- The low rotating speeds, which maintain the flame at the maximum temperature for a longer time (10–20 msec).

Table 2-1
Pollutant Concentrations in Some Gas Turbine Exhausts with Natural Gas Fuel*

Turbine	ppmv Corrected to 15% O_2			g/kWh			% O_2 in the Exhaust Gases
	NO_x	HC	CO	NO_x	HC	CO	
General Electric Frame 3 (6,900 kW)	75	11	21	1.3	0.06	0.20	17
JP 200 (aero-derivative) (14,800 kW)	108	12	73	1.9	0.06	0.71	17
M 3112 R (with recuperator) (8,200 kW)	112	8	13	2.4	0.04	0.03	18
Saturn Solar (780 kW)	52	65	123	1.0	0.33	1.21	17
Rolls Royce Avon 1534 (16,300 kW)	87	26	100				16.5
Spey 900 (14,240 kW)	134	25	56				17
RB 211-24 A (26,300 kW)	161	28	108				16
RB 211-24 C (29,100 kW)	179	15	76				16
General Electric LM 2500-D (13,150 kW)	118						
LM 2000-E (22,000 kW)	185						
LM 5000 (33,400 kW)	267						

* For liquid fuels, the NO_x content is 60% to 90% higher.

Different methods are used to reduce pollutants CO, UHC, and NO_x in gas turbines.

It is difficult to reduce the quantity of NO_x due to the fuel-bounded nitrogen (FBN), but it is possible to reduce the amount formed by the air in the flame with the following methods

Table 2-2
Improved Pollutant Concentrations in Some Gas Turbine Exhausts with Natural Gas Fuel

Turbines	2-cycle or 4-cycle	NEMA Power @ 27°C (kW)	g/kWh		
			NO_x	HC	CO
Natural aspiration					
I.R. KVG 8	4	590	16.7	1.75	0.64
CB GMV 10	2	1,000	34.5	4.77	1.10
CB GMWA 6	2	1,100	18.6	8.55	0.71
Clark BA 8	2	1,185	14.2	8.25	2.50
Turbocharged					
I.R. 412 KVS	4	1,480	21.10	4.16	0.85
Clark TLA 6	2	1,480	12.9	5.67	3.24
CB GMWC 6	2	1,480	14.9	7.35	1.24
I.R. 616 KVT	4	2,960	10.9	3.77	1.41
Clark TCV12	2	2,960	11.05	7.58	4.26
CB 14V250	2	3,555	24.9	6.30	2.00
I.R. 616 KVR	4	4,075	14.75	2.98	1.21
Clark TCV16	2	4,075	6.1	6.59	4.73
CB 16V250	2	4,075	24.0	5.28	2.00

I.R.—Ingersoll Rand
CB—Cooper-Bessemer

- Reduction of flame temperature with water, steam injection, or addition of nitrogen in combustion air or in fuel.
- Combustion modification.
- Shortage of the time during which the burned gases remain at high temperature (freezing of the combustion reaction).
- Utilization of catalytic converter which also reduces the UHC and CO content.

All the methods for gas engines and gas turbines will be examined in Chapters 3 and 4.

INTAKE AIR FILTERS AND SILENCERS

The intake air of reciprocating engines and gas turbines contains dust, which can increase wear on the piston cylinders and sliding parts, and get into the lubricating oil, where it will form an abrasive mixture capable of rapidly destroying bearings and all moving parts. Fine dust (<5 microns), as well as salt contained in the fog of marine atmospheres, is de-

36 Drivers for Rotating Equipment

posited on the air-compressor blades and vanes of gas turbines, where it can lower the compression ratio and consequently the power and efficiency. Larger dust particles (>10 microns) erode turbine vanes and blades and cause protective films to disappear. Cleaning the blades with detergents or by injecting solid particles such as abricotine or wheat may bring a compressor back to its rated performance but will also cause wear.

To avoid such problems, it is necessary to protect heat engines with air filters carefully designed for the size of the dust particles to be removed and the rate of dust removal (Figure 2-12). An engine's rate of wear is affected by dust-particle size, dust concentration, the weight of dust entering the engine, particle hardness, particle velocities in the engine, the manner in which particles impinge on wearing parts, and the engine's materials of construction. The many factors that characterize the dust have been grouped (Table 2-3) so as to identify the various environments in which an engine might be operated. Six classes of environments have also been categorized according to which air filters can be specified (Figure 2-13) (e.g., AC Coarse for Classes III–VI or AC Fine for Classes I and II). Desert environments, which have dust with the characteristics shown in Table 2-4, usually require filters of the Ottawa Sand type or an equivalent. Dust-particle sizes are related to filter type and environmental source in Figure 2-14.

If dust samples from the site of an engine's installation are not available, the air filter can be specified according to the environment using

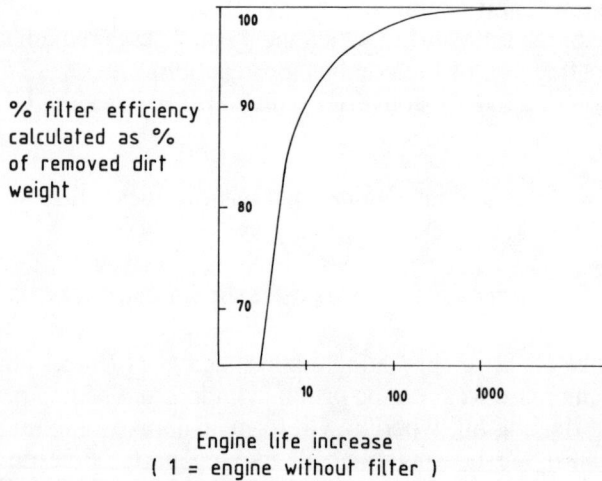

Figure 2-12. The effect of intake filters on the life span of heat engines.

Table 2-3
A Summary of Dust Types and Their Effects on Gas Turbines

	Rural Areas	Coastlines	Large Cities	Industrial Zones	Desert	Tropical	Arctic
Type of dust	Dry, nonerosive	Dry, nonerosive but particles of salt form erosive mixtures	Oily, erosive and corrosive	Oily, erosive and corrosive	Dry, erosive	Nonerosive, fouling	Nonerosive
Concentration	0.01–0.1	0.01–0.1	0.03–0.3	0.1–10	0.1–700	0.01–0.25	0.01–0.25
Particle dimensions (microns)	0.01–3	0.01–3 (salt = 5)	0.01–10	0.1–50 (high near chimneys)	1–500 (high in dust storms)	–.1–10	0.1–10
Temperatures (°C)	–20, +30	–20, +25	–20, +35	–20, +35	–5, +45	–5, +45	–40, +5
Effects on gas turbines	Minimum	Corrosion	Fouling, corrosion	Fouling, erosion, corrosion	Erosion, insect fouling	Fouling	Fouling

38 Drivers for Rotating Equipment

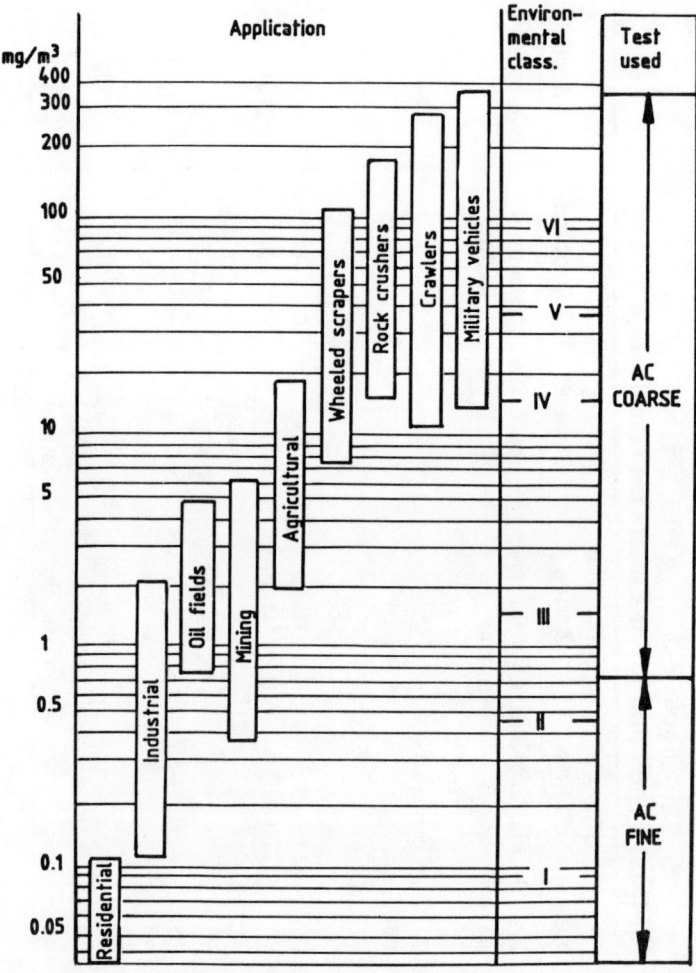

Figure 2-13. Dust contents of ambient air, shown as mg/m^3, vs. the type of environment, the environmental class, and the applicable intake filter. (Courtesy of Donaldson Cy, Inc.)

Table 2-4
Sizes and Amounts of Dust Particles Removed by Standard Intake Filters

Percent by Weight	Particle Dimensions (microns)		
	AC Coarse	AC Fine	Ottawa
25	< 10	< 2	<280
50	< 30	< 7	<380
75	< 60	<20	<420
100	<180	<60	<700

General Characteristics of Heat Engines 39

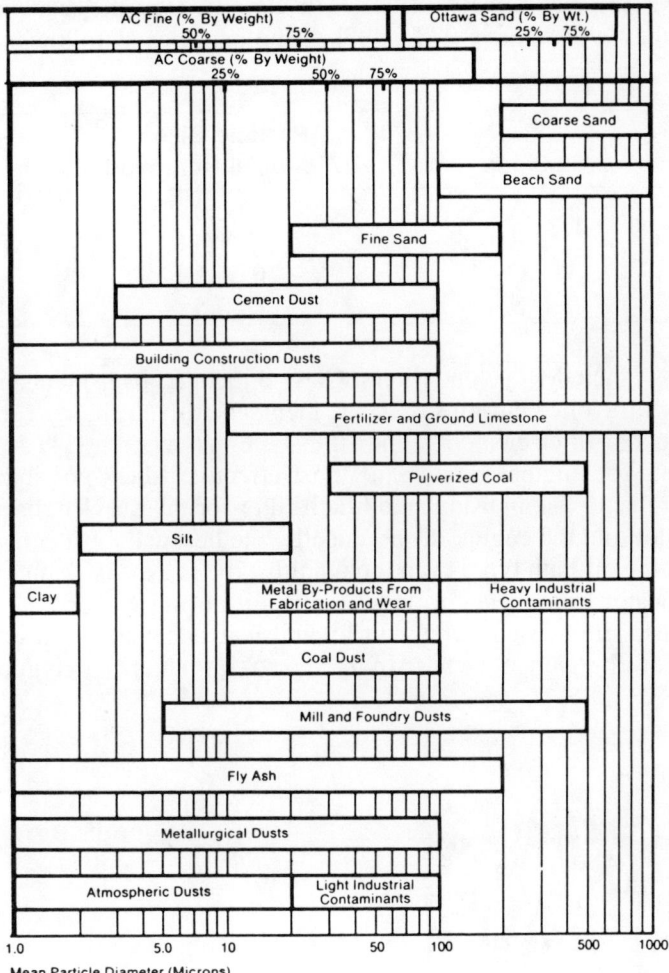

Figure 2-14. Dust particles in ambient air, shown as mean particle size, vs. source and applicable intake filter. (Courtesy of Donaldson Cy, Inc.)

Figure 2-13. Additionally, filters are classified by the size range, by which they will remove 97%–98% of the particles as follows:

Class	Particle Size 97%–98% Removed
A	10–30
B	3–10
C	0.5–3
S	< 0.5

Class A includes cyclone filters; Class B includes bag filters and filters with metallic oiled curtains; Class C includes oil-bath filters, filters with impregnated filter-elements, and filter-type barriers.

Piston-type engines require the most effective filters possible—Class C. There is no risk in using an oil-bath filter for this service, because the oil cannot foul the engine. Consequently, the best and most conventional filter is the oil-bath type, even though this type is expensive and requires an oil heater or a change to less viscous oil during winter.

Because air entering through the cylindrical side walls makes a sudden change of direction at the base of the typical oil-bath filter (Figure 2-15),

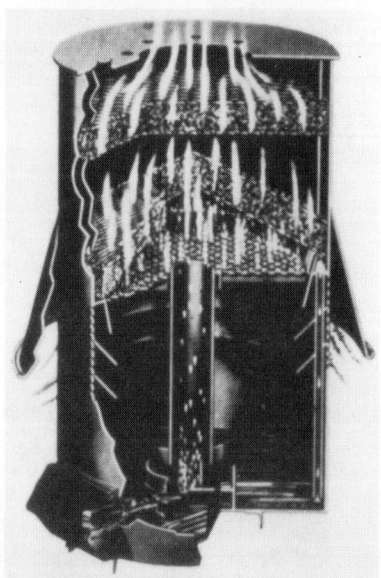

Figure 2-15. American Air Filter Cycoil type W oil-bath air filter.

General Characteristics of Heat Engines 41

it mixes intimately with the oil in the bath at the bottom. Considerable oil is carried upward by the swirling air, and the dust-laden droplets are plastered on the interior walls of the cylinder by centrifugal force, to flow back down to the lower tank. Two mist separators in the upper part of the filter stop oil and dust that have not been removed through centrifuging.

The maintenance of oil filters is limited to periodically changing the oil in the lower tank. If this operation is done frequently, the two mist separators need no special cleaning. Nevertheless, it is possible to dismantle them through the lower section.

The effectiveness of oil-bath filters is 99.91% for particles larger than 3 microns. At 25% of its maximum capacity, a typical oil-bath filter still has an efficiency of 99.65%. The pressure drop is around 100 mm of water (10 mbar).

Another type of filter with very good efficiency, but in Class B, is the filter with a rotating (1 turn in 24 hours) metal curtain, which is impregnated with oil and is cleaned by going through an oil bath at its base (Figure 2-16). Dust that hits the curtain is stuck there, to be washed off in the

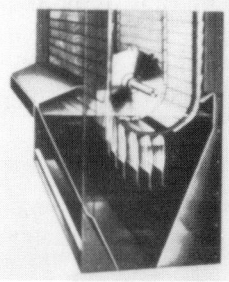

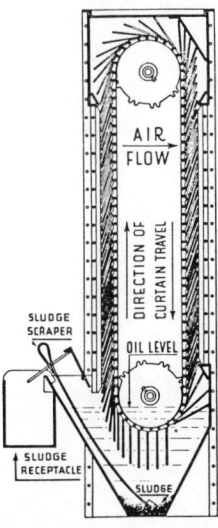

Figure 2-16. American Air Filter multiduty filter.

bath. However, dust of 1–5 microns is not stopped, and can lead to clogging. This filter has the disadvantage of not muffling the engine's intake noise, so that it is necessary to place a silencer between the intake and this filter. Such a silencer, which is usually made of perforated glass wool and a metal sheet has a tendency to clog with oil vapors, and must be purged. It can also catch fire in the case of a flash-back due to poor sealing at the engine's inlet. The pressure drop of this type filter is 0.8–1 mbar (10 mm of water).

Oil-bath and curtain filters can become filled with water in the event of rain or fog, so that their oil baths must be inspected during inclement weather. They also are not effective against salt.

Piston engines may also be equipped with air filters containing filter elements of fiberglass or synthetic fiber impregnated with resin and a viscous fluid. These filter elements are removable (Figure 2-17), and their size is chosen for a low pressure drop on the order of 5 mbar. The air velocity through them must be less than 2 m/sec. They have the advantage of doubling as an intake silencer as well as a filter, but they gradually get plugged, leading to increased pressure drop and reduced power and efficiency from the engine. On a normal site the lifetime of the filter medium is one to two years. The filters must be changed often if the dust content of the air is high or if there is a recycle of oil vapors coming from the engine's vents. Thus it is preferred practice to use oil-bath filters where the air is particularly charged with dust.

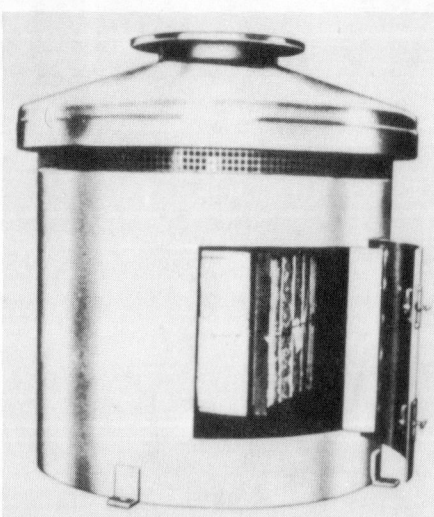

Figure 2-17. Burgess Manning BVV filter with interchangeable elements.

Open-cycle gas turbines require removal of dust with particle diameters above one micron. Below one micron, particles do not have a significant effect; at one to five microns they can clog the air compressor; and over ten microns they cause wear. Accordingly, the filters used for open-cycle gas turbines include a combination of Class A or Class B with a subsequent Class C filter. The Class A filters are generally cyclone filters; the Class B are generally bag filters; and the Class C are those containing filter elements of fiberglass or synthetic fiber.

Cyclone filters operating at a pressure drop of 2.5 mbar (Figures 2-18 and 2-19) have an effectiveness of 95% for particles larger than 10 microns and only 65% for particles of 2.5 microns. Double-cyclone filters

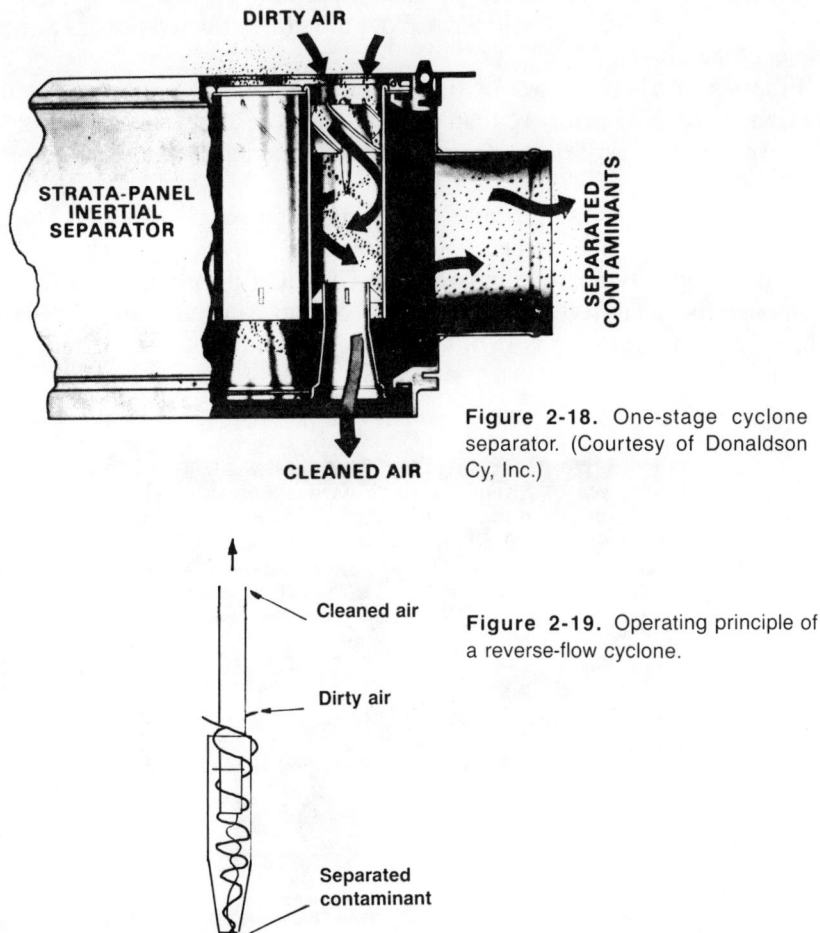

Figure 2-18. One-stage cyclone separator. (Courtesy of Donaldson Cy, Inc.)

Figure 2-19. Operating principle of a reverse-flow cyclone.

44   Drivers for Rotating Equipment

operating at a pressure drop of 5–8 mbar (Figure 2-20) have an effectiveness of 90% for particles of 2.5 microns. The dust is removed by means of a purge with a fan. The air velocity must be 5–10 m/sec for good operation. These filters, which require no maintenance, must be used as prefilters to avoid frequent changes of the elements of that type of Class C filters, whenever the dust content of the air exceeds that of a residential environment (Figure 2-13). For example, air containing 0.2 mg/m^3 of dust corresponds to the atmosphere of a large city and a 25 MW gas turbine operating 10 hours per day in such an atmosphere would be presented with 255.5 kg/yr of dust to eliminate.

Also suited for the same use as cyclones are filters with baffles that cause the air to change direction so that the dust is thrown out by its momentum (Figure 2-21). These have about the same effectiveness as a simple cyclone filter.

Filters with flexible bags of filtering material are also used as prefilters; they are more effective than the cyclone filters but less well adapted for large quantities of dust. They also muffle the intake noise of the turbine.

Class C filters use removable 10–15 cm-thick panels of either fiberglass or synthetic fiber impregnated with resin and a viscous fluid. The pressure drop is typically 5 mbar. Some Class C filters contain elements fashioned like an unrolling curtain of dry or impregnated fiberglass (Figure 2-22). An electric motor rolls out a clean area of filtering material as

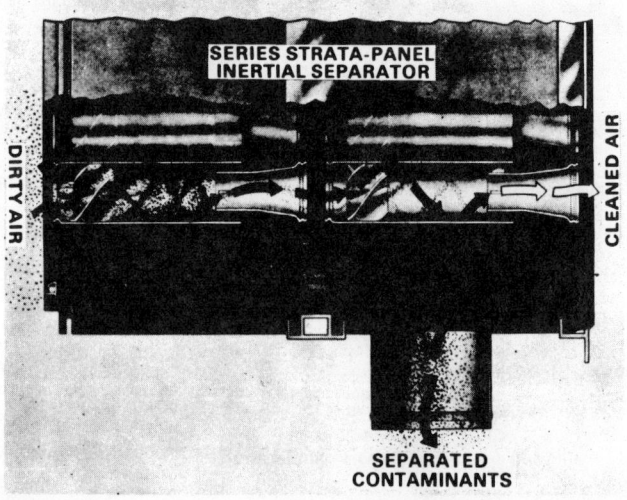

Figure 2-20. Two-stage cyclone separator. (Courtesy of Donaldson Cy, Inc.)

General Characteristics of Heat Engines 45

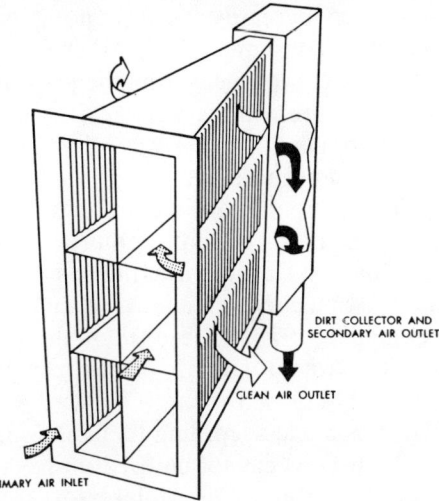

Figure 2-21. Louvered cyclone separator. (Courtesy of American Air Filter.)

Figure 2-22. Roll renewable panel filter. (Courtesy of American Air Filter.)

46 Drivers for Rotating Equipment

soon as the pressure drop increases past a set point due to clogging. This system is excellent, but it consumes a lot of filtering material if a prefilter is not used and if the system for measuring the pressure drop is not operating properly. The pressure drop is 5 mbar. It is important that the air velocity for these filters does not go over 2.5 m/sec, in order to avoid risk of forming ice in humid weather.

Some Class C filters with only one stage and a highly effective paper-type medium have a system for periodic compressed-air injection in a direction opposite to flow for flushing the filtering cartridges. This partial regeneration at the end of a cycle is initiated when the pressure drop goes over a preset level (Figure 2-23). These filters do not require prefilters, the velocity of the air is from 0.9 to 1.5 m/mn through the filter medium; and the pressure drop is 2.5–7.5 mbar.

The choice of filter is made according to Table 2-5. Filters with an oil bath or impregnated curtains cannot be used on gas turbines because of the risk of clogging the turbine's air compressor.

Inlet Silencers

The high frequency sound produced by the axial air compressor of gas turbines must be reduced by silencers made of glass wool sandwiched between perforated metal sheets located between the air filter and the turbine intake (Figure 2-24). The compressor noise is propagated as plane waves and cannot go around obstacles. They are also reflected just as light is. This is the reason sound-absorbing baffles spaced in air passages with elbows are used (Figure 2-25). The average sound is absorbed by baffles of perforated sheet metal filled with glass wool, and its frequency

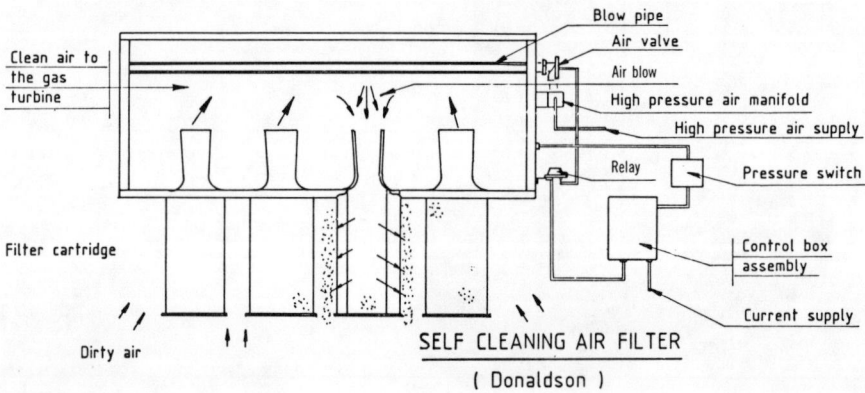

Figure 2-23. Automatically regenerated panel filter. (Courtesy of Donaldson Cy, Inc.)

Table 2-5
Recommended Filters for Different Environments

Environmental Class, concentration, (mg/m³)	Environment	Type of Filter
I, 0.1	Residential. Industrial zones where 95% of the surface is covered and roads are paved.	Simple cyclone filters, bag filters, or if fouling is to be prevented, filter panels.
II, 0.45	Industrial area, with 60% of the ground covered, rainfall is moderate.	Simple cyclone or bag filters, possibly with filter panels to prevent fouling; a cylcone filter alone risks fouling.
III, 1.4	Dusty industrial areas, where 30% of the ground surface is covered, roads are not paved, and rainfall is minor.	Two-stage cyclone, bag filter possibly with a panel filter to prevent fouling; automatically regenerated panel filter.
IV, 14	Agricultural areas (without pasture land) minor rainfall.	One-stage cyclone with a panel filter follow-up, automatically regenerated panel filter.
V, 35	Urban rather than industrial working areas.	Two-stage cyclone with panel filter follow-up; panel filter with automatic regeneration.
VI, 70	Terrain for military maneuvers. Areas near earth-working or earth-moving operations.	Two-stage cyclone, plus bag filters, plus panel filter, or automatically regenerated panel filter.
VII	Coastline, with significant wind and humidity and salt water sprays and mists.	Filter to eliminate water, plus protecting screens followed by a panel filter. Automatically regenerated panel filter.
VIII	Snow with possibilities for freezing rain, fogs, sleet, possibility of icing.	Cyclone type filter, plus a panel filter with de-icing system. Panel filter with automatic regeneration.

Courtesy of Donaldson Cy Inc.

48   Drivers for Rotating Equipment

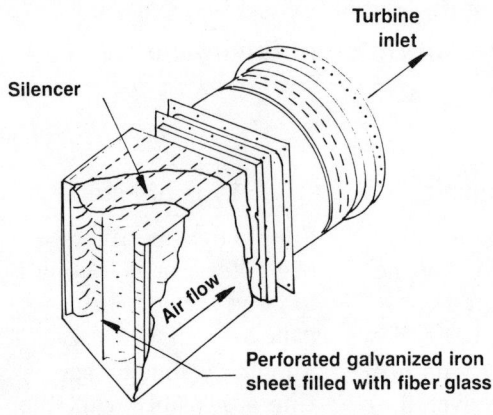

Figure 2-24. Typical gas turbine inlet silencer. (Courtesy of Solar Gas Turbines, Inc.)

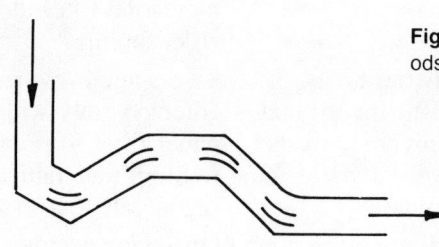

Figure 2-25. Noise-reducing methods for sharp sounds.

can be changed and amplitude reduced with changes in the piping diameters.

Silencers are rarely needed between the filter and suction of gas engines; and when needed they are simple silencers with a length of 10–20 pipe diameters (Figure 2-26).

Exhaust Silencers

The exhaust of an open-cycle gas turbine must be muffled with a one- or two-stage silencer (see Figure 2-27), in order to eliminate the sharp sounds produced by the expansion turbine. The first stage is made up of baffles of perforated metal sheet filled with glass wool, possibly arranged in a zigzag. The second stage is an empty duct with perforated sheet-metal walls and glass wool behind them. These silencers' effectiveness increases with their length; and a heat-recovery unit on the exhaust makes an excellent silencer.

General Characteristics of Heat Engines 49

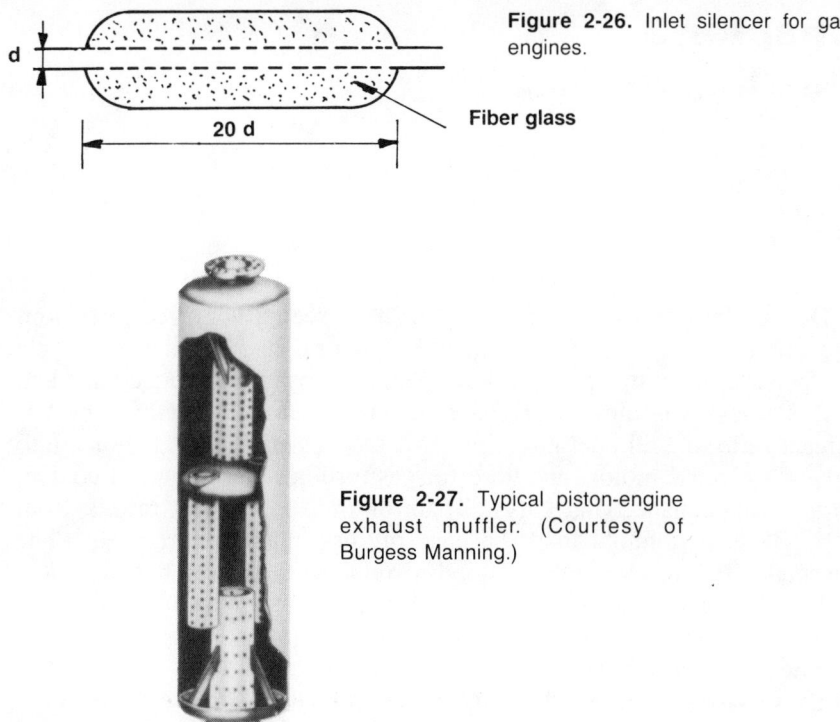

Figure 2-26. Inlet silencer for gas engines.

Fiber glass

Figure 2-27. Typical piston-engine exhaust muffler. (Courtesy of Burgess Manning.)

The exhaust noise spectrum of a piston engine has two different components:

- The first one in the low frequencies is directly related to the rpm and to the combustion noise.
- The second one in the high frequencies is generated by the turbocharger.

The silencer must be able to attenuate both the low and the high frequencies, but particularly the low frequencies which propagate for long distances and are still present at several hundred meters. The high frequencies are attenuated more rapidly by air molecular absorption.

Usually the silencer combines a "low pass filter" (usually called a "resonator") designed to absorb the noise energy in the fundamental and the first harmonics for the full range of operating conditions and an absorptive section with acoustical pack material (mineral wool) to absorb high frequency noise generated by the turbocharger.

Chapter 3
Gas Engines

During the last 40 years gas engines have seen a very great development related to the ever expanding use of natural gas. Whether two-cycle or four-cycle, with spark ignition or ignition by gas oil injection (dual fuel engines), gas engines have acquired a ruggedness such that their on-stream ratio to total operating time now reaches 96%-98% compared to 99% for electric motors and 98% for gas turbines (assuming, of course, proper maintenance and care). Their thermal efficiency ranges from 30%-38% at nominal load, because of supercharging processes. The thermal efficiencies are understood to refer to the lower heating value (LHV) and take into account driving accessories, such as lube-oil pumps and water pumps. Also, in order to reduce investment costs, the unit capacities have been carried up to 12,000 kW.

These machines are used for driving reciprocating compressors in integrated trains, but they are also suitable for driving centrifugal compressors, pumps, and alternators by means of speed-multiplying couplings. For their pipeline use in driving compressors and alternators, they are always spark-ignited for reasons of convenience and operation cost even though the oil-ignited dual-fuel system is less sensitive to variations in the air-fuel ratio and its ignition is more reliable.

Natural gas, which contains a lot of methane, is the best hydrocarbon for combustion in engines. It has the least tendency for auto-detonation (the octane number of methane is about 140). It should be noted also that butane and propane are also excellent fuels with good anti-detonating properties, whereas coke oven gas, which has a high hydrogen content, tends to auto-ignite at compression ratios over 6, so that obtained efficiencies are low.

Gas engines can have 4–20 cylinders with a power per cylinder ranging from 75–600 kW. Their nominal speeds range from 300 to 1,200 rpm.

Figure 3-1 outlines the four stages of a four-cycle engine, as well as the time distribution between the four stages. This diagram corresponds to two revolutions of the crankshaft and includes four stages, of which only one is driving. High-speed engines generally have the intake opening more advanced and the exhaust closing less delayed. Low-speed engines have the intake opening less advanced and the exhaust closing less delayed.

The advance of the intake opening and delay of the intake closing are controlled so that the cylinder can be completely filled with air, with allowances for the intake pressure drop. The control of ignition advance is determined in such a way that the maximum pressure point is situated after the top dead center, and the retreating movement of the piston has just started.

The control of the exhaust valve opening is obtained experimentally to have the cylinder pressure at the end of expansion fall in the neighborhood of atmospheric pressure. The delay of the exhaust-valve closing is controlled so that all the burned gases are well purged from the cylinder.

With supercharged engines, it can be seen (Figure 3-1) that the advance of the intake opening and delay of the exhaust closing are increased relative to the standard motor, so that the cylinder will be flushed for total removal of burned gases, as well as cooling of the cylinder walls and the head and top of the piston. If the engine is supercharged at 2–3 bar abs. pressure, as is the Nordberg Supair-thermal engine for example, the closing of the intake valve is still positioned before the bottom dead center at an instant determined as a function of the load and of the possible air-temperature lowering due to expansion between the intake-valve closing and the bottom dead center.

Figure 3-2 shows the stages of a two-cycle engine and its time distribution. The diagram corresponds to one revolution of the crankshaft during which it is necessary to accomplish the four steps that take place on a four-cycle engine.

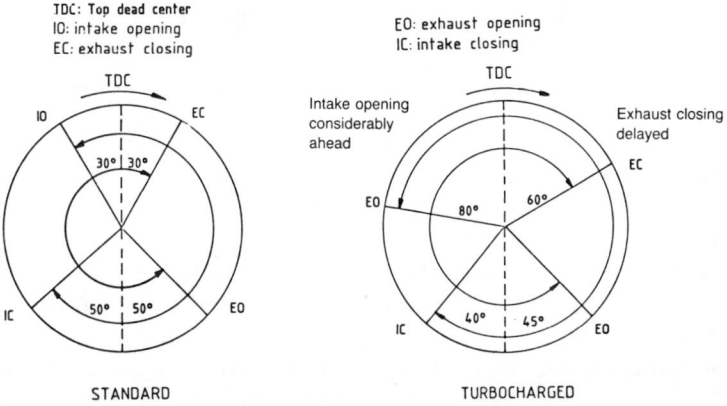

Figure 3-1. The operation and timing of a four-cycle gas engine.

ASVO: air starting valve opening
EO : exhaust port opening
IO : intake port opening
ASVC: air starting valve closing

FVO : fuel valve opening
IC : intake port closing
EC : exhaust port closing
FVC : fuel valve closing

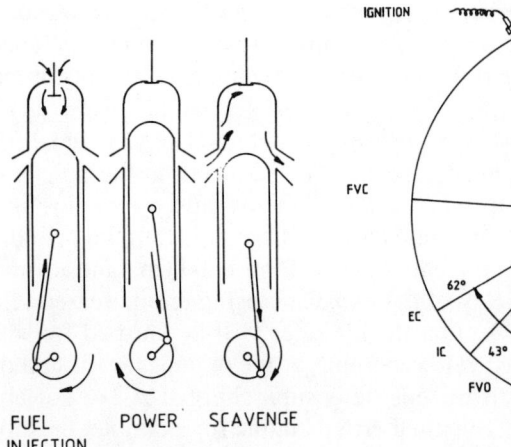
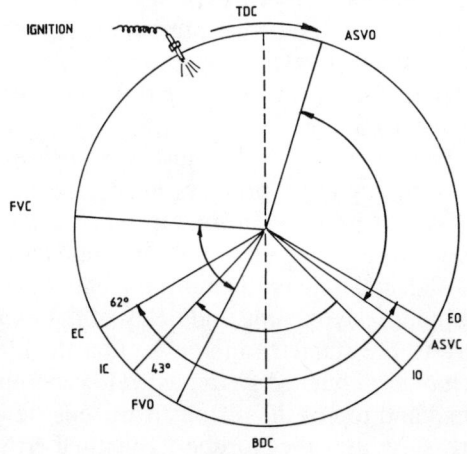

Exhaust 62° BBDC to 60° ABDC - Starting air 15 to 120° ATDC
Intake 43° BBDC to 43° ABDC - Fuel injection 33 to 93° ABDC

Figure 3-2. The operation and timing of a two-cycle gas engine.

Thus there is one power stroke for each turn of the crankshaft, whereas the four-cycle engine has one power stroke for each two turns of the crankshaft. Consequently, in order to obtain the same power, the four-cycle engine must be sized larger. On the other hand, the thermal load carried by the cylinder walls, piston head, and cylinder head of the four-cycle engine is less than that of the two-cycle; in the two-cycle engine the air scavenging the burned gases has the important second function of cooling the piston head, cylinder walls, and cylinder head.

Some two-cycle engines are provided with exhaust valves, which are controlled by a cam shaft and usually require valve rockers (Figure 3-3). On large engines such exhaust valves and their control mechanism require more maintenance than do exhaust ports in the cylinder walls.

Fuel gas is injected to the cylinder under a pressure of about 10 bar. The injection valve opens at the beginning of compression and (in the case of two-cycles) after the intake ports are closed; it closes before the cylinder pressure goes over the gas injection pressure. Figure 3-4 shows the pressure-volume diagrams of a four-cycle and a two-cycle engine.

The horizontal lower loop of the four-cycle diagram plots the intake and exhaust strokes; its area represents the work of scavenging the cylinder (Figure 3-5).

This scavenging loop does not appear on the two-cycle diagram, but the corresponding work is still lost and is furnished by the scavenging-air blower or the supercharger's turboblower, which is necessary to two-cycle engines (Figure 3-6). Two-cycle engines can have higher powers than four-cycle engines, whose capacity is limited by the size of their valves.

Those engines with supercharges, either four- or two-cycle, are the only ones that can have the rate of air flow regulated without energy loss,

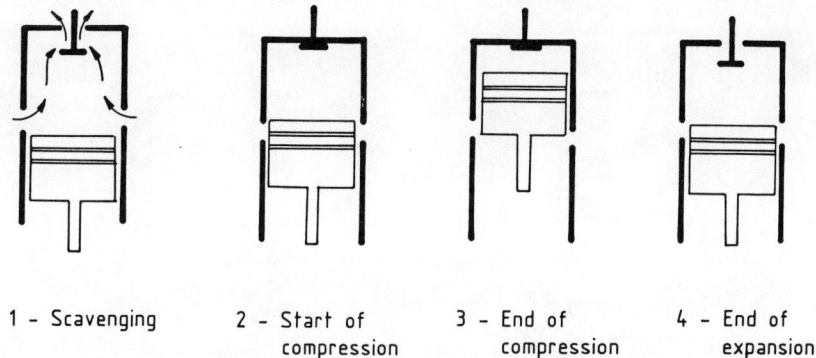

1 – Scavenging   2 – Start of compression   3 – End of compression   4 – End of expansion

Figure 3-3. The operation of a two-cycle gas engine with an exhaust valve.

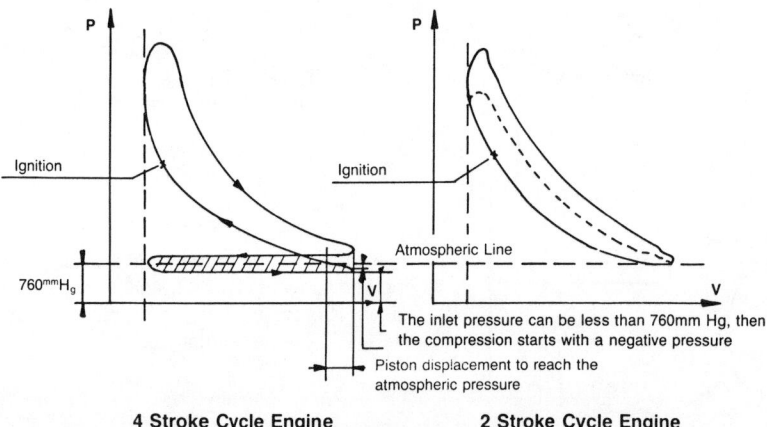

Figure 3-4. Pressure-volume diagrams for four-cycle and two-cycle gas engines.

54 Drivers for Rotating Equipment

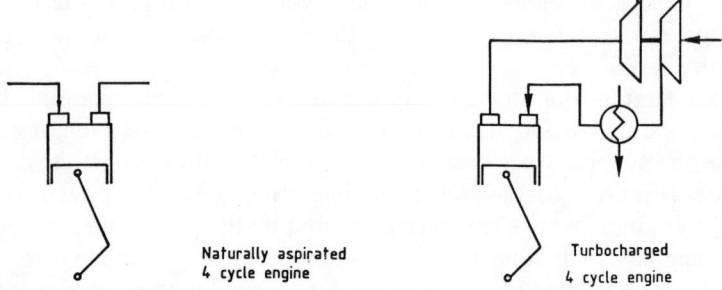

Figure 3-5. Scavenging the cylinders of naturally aspirated and turbo-charged four-cycle engines.

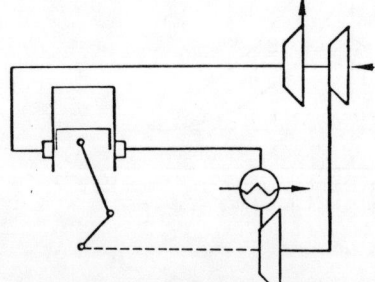

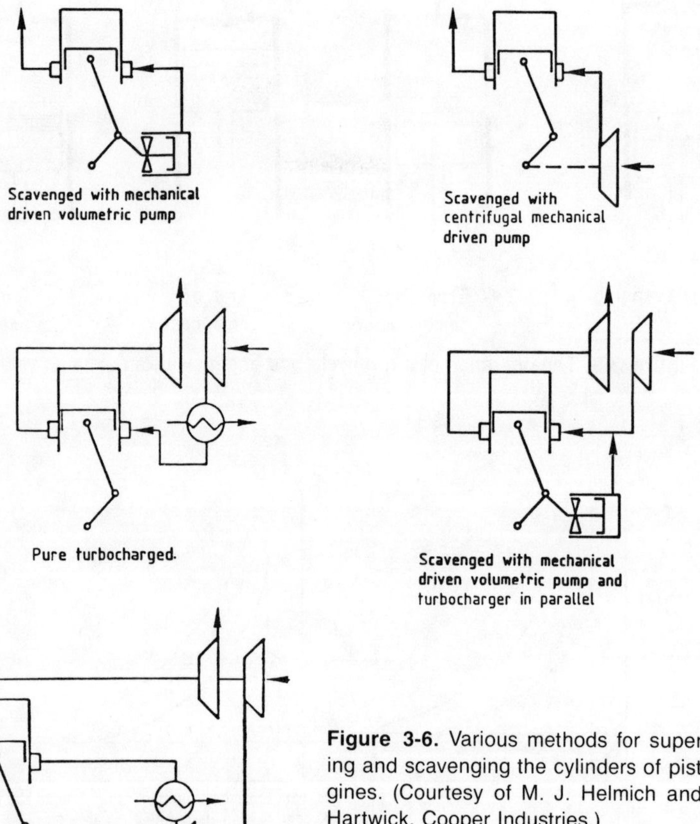

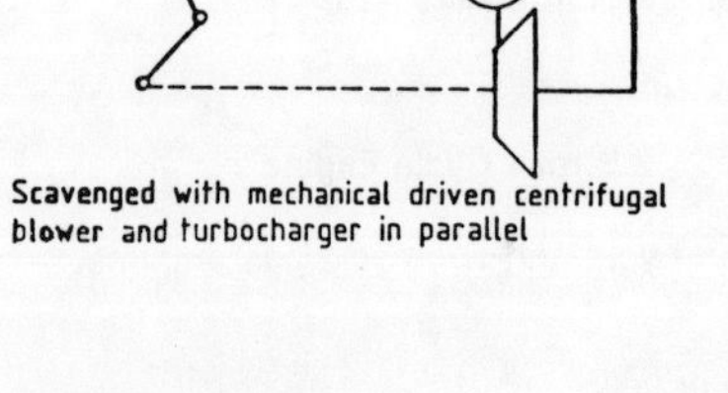

Figure 3-6. Various methods for supercharging and scavenging the cylinders of piston engines. (Courtesy of M. J. Helmich and W. F. Hartwick, Cooper Industries.)

by means of a butterfly valve at the turbocharger's discharge or a bypass around the turbine (Figure 3-6). On the four-cycle engines with natural aspiration (Figure 3-5) or two-cycles scavenged with a positive displacement blower (Figure 3-6), a throttling butterfly valve must be installed in the intake pipe of the four-cycle or on the blower for the two-cycle.

The large gas engines are started with air directly injected in the cylinders or with a rotating starter on the flywheel.

CHARACTERISTICS OF GAS ENGINES

Gas engines are characterized by their:

- Power.
- Rotating speed.
- Overall thermal efficiency, which is the ratio between the effective work delivered by the shaft at constant operating conditions during a number of full cycles to the total energy liberated by perfect combustion (LHV) of the fuel consumed during the same number of cycles. The thermal efficiency is calculated with oil and water pumps driven by the engine.
- Mean effective pressure, which is the pressure caused by that piston displacement representing the work of a cycle.

The mean effective pressure and power are related to each other as follows:

- For four-cycle engines:

$$W = \frac{N.V.\ MEP}{1,200}$$

- For two-cycle engines:

$$W = \frac{N.V.\ MEP}{600}$$

where W = power, kW
       N = rotating speed, rpm
       V = piston displacement, liters
       MEP = mean effective pressure, bar

The MEP also characterizes the torque delivered by an engine. Typical values are:

- For non-supercharged engines: 6 bar (90 psi)
- For supercharged two-cycle engines: 7.5–9 bar (110–130 psi)
- For supercharged four-cycle engines: 9.5–11.5 bar (140–175 psi)

THERMODYNAMIC CYCLE

Figure 3-7 shows the theoretical thermodynamic cycle of a gas engine, with volumes V and v, between which the mass of gas expands, and the maximum pressure, P, which is limited by auto-ignition. The combustion and energy liberation occur in large part at constant volume, as in the Beau de Rochas cycle, and in small part at constant pressure, as in the diesel cycle (Figure 2-3). This explains the origin of the name "mixed cycle" or Sabathe cycle. The diagram approaches the Beau de Rochas cycle more closely as the fuel's combustion rate increases and the ignition occurs well before the top dead center, which is not possible with a diesel cycle, where auto-ignition is caused by raising the air pressure.

Thermal Efficiency

The thermal efficiency varies as an exponential function of the volumetric ratio, with increasing volumetric ratios bringing decreasing improvements in efficiency. Also, the risks of detonation increase with an increase in volumetric ratio, so that contemporary ratios do not go over about 9 for a supercharged engine and about 7 for a four-cycle engine with natural aspiration.

The thermal efficiency of an engine also depends on its air-gas ratio, its method of operation, and its ignition control.

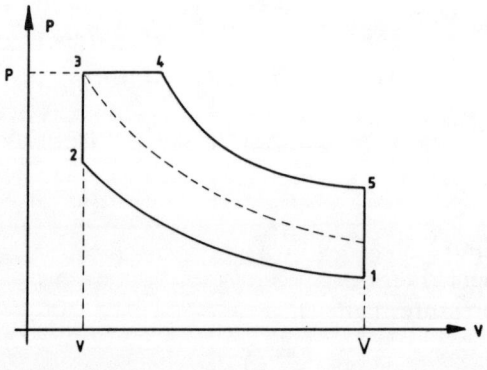

Figure 3-7. Theoretical cycle diagram for a gas engine.

Air-Gas Ratio

Theoretically, to have the best thermal efficiency, the fuel-air equivalence ratio (richness) should be 1.0 to achieve complete adiabatic combustion as well as the maximum temperature, (in fact the maximum temperature is reached with a richness of 0.99). The auto-ignition temperature for natural gas and air in stoichiometric ratio of 1 to 10 in volume and 1 to 17 by weight, and at atmospheric pressure, is 720°C and decreases when pressure increases. Thus the volumetric ratio is limited to 7 to avoid pre-ignition and detonation (the stoichiometric ratio by weight is 1 to 17 for methane, 1 to 16 for ethane, and 1 to 15.6 for propane). Thus for a volumetric ratio of 7, an adiabatic compressibility factor of 1.35, supercharged air cooled to 50°C and diluted by 12% of burned gases at 315°C (two-cycle engine), the temperature at the beginning of compression is 85°C and at the end of compression is 435°C, which is the temperature of auto-ignition. To extend this detonation limit and increase the compression ratio and in consequence the thermal efficiency, the air-gas mixture must have a richness lower than 1.0, and an excess of air, so that the ignition no longer occurs in the normal way.

For a four-cycle engine with natural aspiration, the volume ratio is limited to 7, and the air-gas mixture has a richness a little below 1.0 (0.93 to 0.95). If the richness is allowed to exceed 1.0, the engine enters the zone of CO_2 formation, and the CO content of the exhaust increases rapidly (Figure 2-10). For supercharged four- or two-cycle engines the combustion air is only limited by the capacity of the blower and the fuel air equivalence ratio is definitely lower than 1 (0.8 to 0.9), which corresponds not only to a good ignition stability but also to the maximum NO_x emission level. The lower limit of richness for correct ignition is 0.70, and it must not be exceeded.

In two-cycle engines, where the supercharged air must scavenge the burned gases as well as cool the piston and cylinder, the amount of scavenging air is 140%–150% of the theoretical stoichiometric mixture. This would lead to a minimum richness of 0.67 in the impossible event that all the scavenging air stayed in the cylinder; in actual operation, it leads to a richness in the cylinder of 0.80.

Allowing that the reduction in richness extends the detonation limit, it is possible to use a higher volume ratio and achieve an increased efficiency; but the combustion temperature is also reduced, and the best efficiency for supercharged engines evolves at a volume ratio of 9 and a richness of 0.70–0.75 (Figure 3-8).

On engines with carburetors, where the carburetor is the only control of the air-gas ratio, the richness will vary as a function of the load. At maximum load, there is a risk of detonation if the quality of the gas varies to include more heavy products, such as ethane, propane, and butane. At

58 Drivers for Rotating Equipment

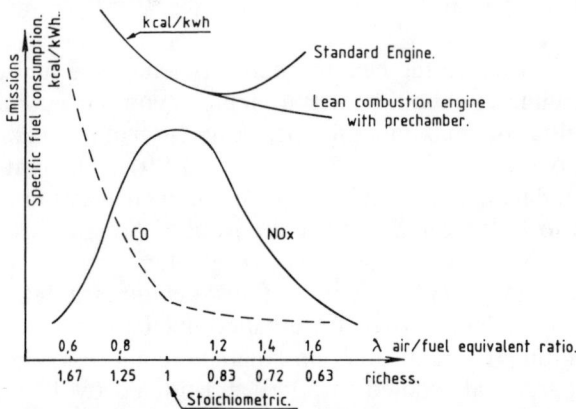

Figure 3-8. Pollutant emissions and specific fuel consumption function of richness.

partial load, there is the risk of failure to ignite, or missing, because the mixture is too poor.

Thus it is important to maintain the air-gas ratio at its optimum value, no matter what the ambient temperature and load, by controlling the flow of air, with the air pressure varying linearly with the pressure of the gas and the gas pressure varying linearly with the torque. Under such conditions the thermal efficiency remains optimal without detonating or misfiring. In order to extend the limits of detonation (and also reduce NO_x), it is necessary to cool the air for combustion.

Efficiency declines with declining load; it declines faster with declining torque than with declining speed. This is due to the fact that at partial load the pressure in the cylinder is lower (it varies like torque) and the ignition has less regularity with some misfiring. The explosion pressures vary greatly. Thus at full load for an explosion pressure of 67 bar the average variations are 4%, and at 25% load the average explosion pressure is 31 bar with variations of 20%. Thus when the load is reduced it is important to maintain the torque and to reduce the speed, rather than the reverse. This is possible when the engine drives a reciprocating compressor.

Ignition

The ignition must be controlled so that the maximum pressure occurs a little after top dead center of the piston stroke, and the exhaust temperature is the lowest possible, i.e., the gas expansion is the greatest. This means that the ignition must be advanced when the torques increase and be retarded when the speeds are reduced; it must be advanced when the air temperatures go down and retarded when they go up.

The turbo-blower must furnish enough air; if it is throttled, the ignition must be retarded to increase the heat in the exhaust gases by reducing the thermal efficiency. If the combustion continues well past top dead center the exhaust gases tend to be hotter. This explains the importance of having a turbo-blower in good clean conditions.

The spark-plug gapping also conditions the thermal efficiency (Figure 3-9).

Energy Balance

The energy of the fuel is dissipated as follows:

	Supercharged Gas Engine (%)	Automotive Engine (%) (for comparison)
Work on the shaft	37	25
Power to auxiliaries	2–3	included in losses
Mechanical losses	5	15–20
Cooling mechanical parts	20	15–20
Exhaust	35	30–40
	100	100

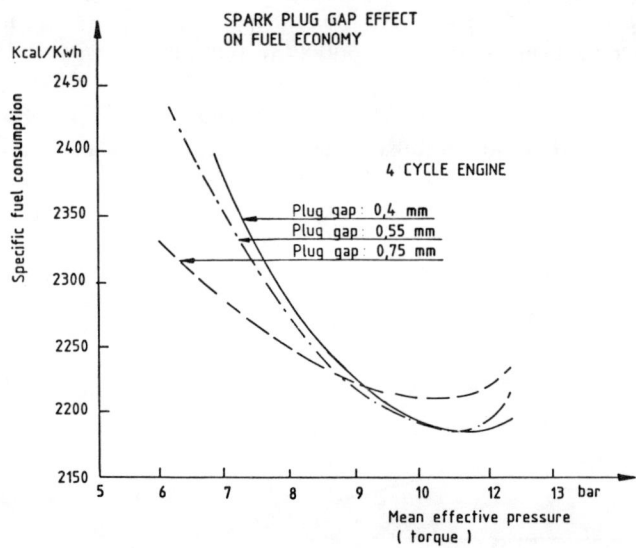

Figure 3-9. Effects of spark-plug advance on the operating characteristics of a four-cycle engine. (Courtesy of M. J. Helmich and W. F. Hartwick, Cooper Industries.)

Detonation

Combustion is normally a flame deflagration. The flame spreads through thermal conduction; the pressure is uniform in the mixture; and the flame velocity is on the order of tens of meters per second. However, the combustion can become detonating, when the flame is propagated by a pressure wave with an abrupt front representing adiabatic compression, and the velocity of the combustion front is on the order of 1,000 meters/sec. There is no stable intermediate between these two types of propagation. Combustion becomes detonating when part of the fuel-air mixture is carried to a sufficiently high pressure or temperature. Fuels are more or less inert to detonation, with their resistance characterized by their octane number.

Detonation is produced in gas engines under five conditions, as follows:

1. The intake pressure is increased, as for example by supercharging, which favors detonation.
2. The intake temperature is increased.
3. The ignition is too much advanced, and the maximum pressure is increased because the mixture is totally burned before the piston reaches top dead center (Figure 3-10).
4. There are hot spots on the cylinder that are insufficiently and irregularly cooled, and which contain roughness or sharp angles.
5. The detonating tendencies of the fuel are more than those for which the engine is being operated, as for example, when there are higher concentrations of heavy products in the gas.

The phenomenon of detonation limits not only the compression ratio of an engine but also the combustion temperature and mean effective pres-

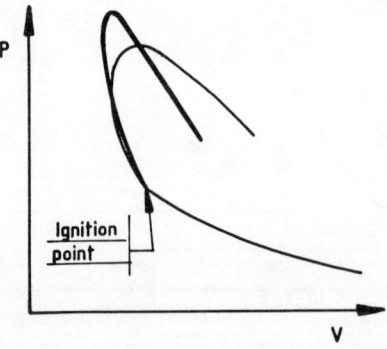

Figure 3-10. Effect of advanced ignition on the maximum cylinder pressure.

sure on which the rated power depends. Detonation is revealed by valve chattering or knocking, abnormal heating, loss of power, a reduction in efficiency, and more rapid wear of the piston rings due to abnormal pressures. Also, hot points can cause local melting of the piston or cylinder head. Nevertheless, the efficiency of an engine is highest when it operates on the verge of detonation.

A gas engine's resistance to detonation is related to its NBN (normal butane number) index, which is the moles of n-butane that can be added to 100 moles of pure methane before detonation occurs at an air-to-gas richness corresponding to optimum efficiency.

Figure 3-11 shows the effect of ignition advance on the volumetric ratio, specific consumption, and NBN index for the same engine. The different volumetric ratios correspond to different cylinder heads.

Cylinder heads with a combustion prechamber for lowering NO_x also raise the NBN index by 2.

Correct matching of the engine to the fuel is important. The volumetric ratio must be adapted for a particular fuel gas defined by its NBN and optimized to provide minimum fuel rate:

1. Excess timing retard with higher compression ratio results in fuel penalty.

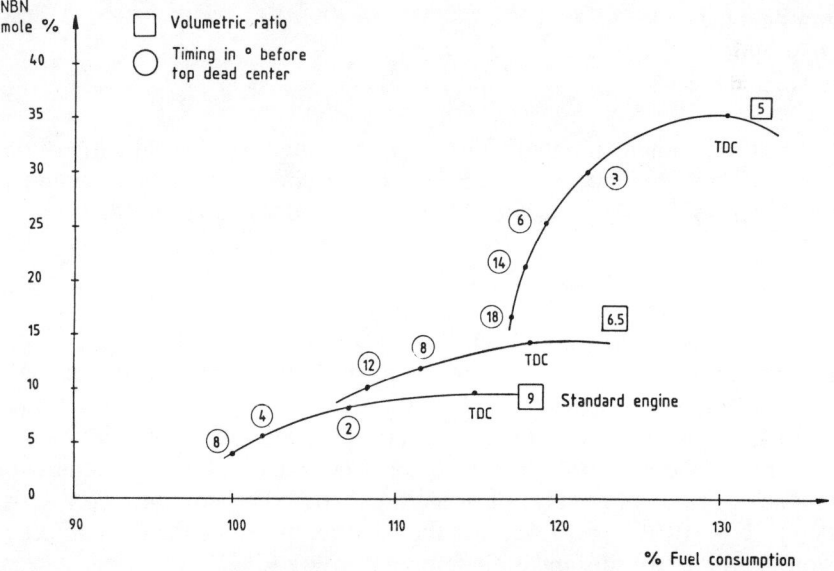

Figure 3-11. Influence on ignition advance, degrees before top dead center, on the operation of a specific gas engine. (Courtesy of Cooper Industries.)

2. Excessive compression ratio reduction adversely affects fuel consumption and increases heat rejection requirements for jacket water and lube oil.

Another procedure, the hydrogen-methane, is used for evaluation of antiknocking properties of gaseous fuels. The methane number is 100 for pure methane and 0 for hydrogen, and the methane number of a gaseous fuel is determined in a similar manner as the octane number for liquid fuels.

The gaseous fuels can be classified:

methane number		
	> 100	excellent
	85 to 100	good ⎫ natural gas
	70 to 85	easy to use ⎭
	55 to 70	difficult to use—refinery gas
	< 55	very difficult to use—coke oven gas

To compare with liquid fuels the octane numbers of different pure hydrocarbon gaseous fuels are:

- Methane—130
- Ethane—112
- Propane—110
- n-butane—95
- n-pentane—60
- Isobutane—100
- Isopentane—92

The final combustion temperature in a gas engine is theoretically of the order of 2400°C, with pure methane fuel, a richness of 1.0, and complete combustion. This can be calculated from standard data, as follows:

$$CH_4 + 2(O_2 + 3.76N_2) \rightarrow CO_2 + 2H_2O + 7.52N_2$$

According to this equation, each 1 m^3 of CH_4 with an LHV of 8,580 kcal/nm^3 requires 2(4.76) = 9.52 nm^3 of air, or each kg of CH_4 with an LHV of 11,989 kcal/kg requires 17.2 kg of air. The specific heat at constant volume, c_v, for the burned gases is 0.19 kcal/kg at 15°C, and $\gamma = c_p/c_v$ = 1.35 for the mixture of air, methane, and burned gases. Assuming an engine with a volume ratio of 9 and adiabatic compression, $(V_1/V_2)^\gamma$ = $(9/1)^{1.35}$ = (P_2/P_1) = 19.42; and the final temperature for the compression is 349°C for an intake air temperature of 15°C.

For 290.76 g of exhaust gas (1 mole CO_2 + 2 moles H_2O + 7.52 moles N_2) the enthalpies in kcal/mole are:

- At 400°C

  $CO_2 = 4.16$; $H_2O = 3.34$; $N_2 = 2.84$:

  $CO_2 + 2\ H_2O + 7.52\ N_2 = 32.17$

- At 2400°C

  $CO_2\ (+\ 10\ N_2) = 61$; $H_2O\ (+\ 5\ N_2) = 43.75$; $N_2 = 19.47$:

  $CO_2 + 2\ H_2O + 7.52\ N_2 = 294.91$

  Difference $(294.91 - 32.17) = 262.74$

For the enthalpies, the dissociation at 2400°C of the CO_2 and H_2O molecules is taken into account. The dissociation is variable with inert quantities; CO_2 and H_2O are considered as inert to calculate the enthalpy of each other. The specific heat at constant pressure from 400°C to 2400°C is $262.74/(2{,}000 \times 290.76) = 0.000452$ kcal/g and the average constant-volume specific heat of the $17.2 + 1.0 = 18.2$ kg of combustion gases is calculated as 0.335 kcal/kg-°C $= 0.452/1.35$, so that the temperature rise, ΔT, is:

$$\Delta T = \frac{11{,}989\ \text{kcal/kg}}{18.2\ \text{kg} \times 0.335} = 1976°C$$

And the maximum temperature in the cylinder is:

$1976 + 349 = 2325°C$

Actually, the maximum temperature will be lower because the fuel-to-air ratio is much lower than the fuel-air equivalence ratio of 1.0, in order to avoid detonation.

With a supercharged engine, the air temperature is 55°C at the outlet of the air cooler; and with a two-cycle engine it is as high as 85°C because of admixture with 10%–12% burned gases at 300°–320°C. Thus the compressed air temperature is about 550°C, and the temperature in the cylinder reaches about 2100°C at the beginning of combustion up to a maximum of 2200°C. Generally, the maximum temperature falls between 1950–2220°C, depending on whether the engines are supercharged. The phenomenon of detonation increases the maximum combustion temperature more than 150°C.

Auto-Ignition (Pre-Ignition)

The phenomenon of detonation can cause localized temperatures high enough to melt the metal, and in automotive engines detonation can cause the perforation of an aluminum piston.

Another phenomenon more dangerous than detonation is auto-ignition in advance of the normal spark-plug ignition (Figure 3-12). Like detonation, this can cause deterioration of the cylinder head, rings, and pistons due to the excess pressure. Whereas detonation can be very easily identified as knocking or valve chattering, auto-ignition can be detected only by an increase of temperature in the cylinder head or the exhaust gases, or by the continued running of the engine after the ignition is turned off. Auto-ignition doubles the surface temperature of the pistons and cylinder head, from about 170°C to about 350°C.

Auto-ignition generally results from one or more of three causes:

1. Too much ash in the oil. The ash collects in a mass that heats up and creates hot spots.
2. Poorly fitting spark-plugs.
3. A hot spot, which will generally be located on a spark-plug or a gas or air injection valve. Because the spark plugs and valves are fixed on the cylinder head, their only means of heat removal is by transfer to the cylinder, and if they are not tight, transfer is poor and they heat up (Figure 3-13). In order to assure the important good contact between spark-plug threads and the cylinder, a cold spark-plug should be mounted in a cold engine, or the cold spark-plug should be hand-mounted in a hot engine and allowed to warm up before loosening and retightening.

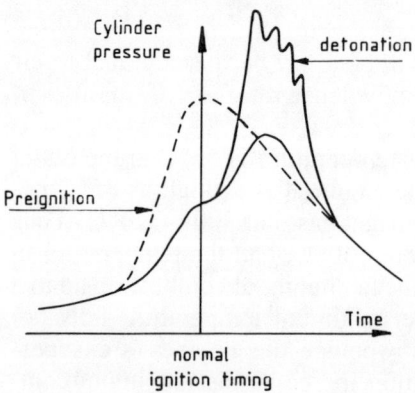

Figure 3-12. The effects on cylinder pressure of preignition, spark-plug ignition, and detonation.

Gas Engines   65

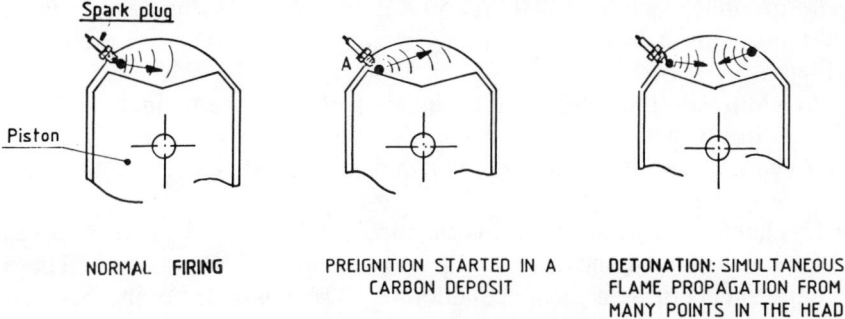

NORMAL FIRING   PREIGNITION STARTED IN A   DETONATION: SIMULTANEOUS
                 CARBON DEPOSIT   FLAME PROPAGATION FROM
                                            MANY POINTS IN THE HEAD

Figure 3-13. Causes of pre-ignition and detonation.

NO_x Emissions

As indicated in Chapter 2, a lean mixture of gas-to-air below the stoichiometric increases the formation of NO_x. Four-cycle engines with natural aspiration operate on lean mixtures close to the stoichiometric. If their richness is increased, the NO_x content is reduced, but the CO content is increased enormously as soon as the richness exceeds 1.0 (Figure 2-11). Therefore these engines achieve a minimum pollution with a mixture that is a little richer than 1.0 while accepting an increase in the CO content and a reduction in efficiency.

The NO_x reduction is 10% to 15% with a specific consumption increase of 3%–4%.

With supercharged engines, the best operating point with respect to efficiency and the volume ratio which is a function of detonation is a mixture with a fuel-to-air ratio slightly below stoichiometric, but which also corresponds to the maximum NO_x. The fuel-to-air ratio for the optimum volumetric ratio of 9 is 0.8. Because of the supercharging, it is easy to have a lean mixture and considerably reduce both the CO and NO_x content of the gases without an unacceptable increase in unburned hydrocarbons. The problem becomes one of stable operation without ignition failures. The accepted fuel-air equivalence ratio for good ignition is a maximum variation of 6 bar between the pressure-crests of the explosions corresponding to 9% of the average explosion pressure; a variation of 8 bar (12% of the average explosion pressure) is unacceptable.
unacceptable.

In a normal supercharged engine the richness varies between 0.8 and 0.9, and the final compression temperature is around 550°C (820 K); combustion heats the gases to about 2100°C (2370 K) with a peak temperature that reaches 2180°C (2450 K); and NO_x is formed at a rate of 700 ppmv/ms and even 1,100 ppmv/ms in the hottest parts of the flame (Figure 2-10).

In addition to the adjustment of the air-fuel ratio, the richness increase for naturally aspirated engines or the maximum decrease for supercharged engines, the following NO_x reduction methods can be used:

- Mechanical adjustments in overlapping the intake and exhaust valves for four-cycle engines and modification of the volumetric ratio. These adjustments bring a small reduction of NO_x emissions but increase the specific fuel consumption
- Timing retard
- Combustion air cooling
- Modification of the combustion for supercharged engines
- Catalytic converter on exhaust gas

Timing retard brings a reduction in NO_x emission but increases the specific fuel consumption. Figure 3-14 is approximative and varies slightly with the richness and the mean effective pressure of the engine. By combustion air cooling, all other parameters remaining about the same, the NO_x formation is reduced to an average of 200 ppmv/ms (300 ppmv/ms peak), as the flame temperature is reduced from 2180°C to 2000°C (Figure 2-10). Thus in winter the NO_x emission is lower.

The modification of the combustion is only possible with supercharged engines where combustion air is available as needed. The combustion can be stratified or lean.

1. If the cylinder head is formed to provide two combustions (a small rich chamber at 1.2 stoichiometric with a flame temperature of 2320°C and a large lean chamber at 0.72 stoichiometric for the flame at 2000°C), the NO_x can be held to about 300 ppmv/ms in the lean chamber and 330 ppmv/ms in the rich.

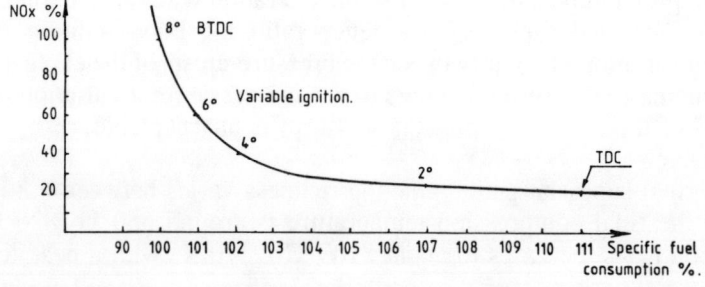

Figure 3-14. NO_x emissions and specific fuel consumption function of timing.

2. If lean combustion is carried out with a fuel-to-air mix of 0.68 stoichiometric, the flame temperature drops to 1900°C and NO_x formation (Figure 2-10) is 90 ppm/ms with a peak of 200 ppm/ms; and at the ratio of 0.62 the minimum NO_x is obtained. The lower the richness is, the lower is the NO_x formation, which causes problems in igniting and maintaining the combustion. One way of solving this problem is to use one or two small precombustion chambers that replace one or two spark plugs in wells, as is commonly done in diesel engines.

In the operation of these cylinders (Figures 3-15 and 3-16), the precombustion chamber's gas injection valve stays open during the course of compression, as long as the pressure in the cylinder and the prechamber is lower than the pressure of the precombustion gas; and the gas enters the prechamber. After this valve closes, compression continues, and combustion gas is injected into the combustion chamber at a pressure higher than the prechamber's gas by the gas injection valve. The spark plug ignites the prechamber's rich mixture, and there is a jet of gas flame out through the orifice of the prechamber to ignite the lean mixture in the combustion chamber and maintain combustion. The combustion time is shorter than a normal spark-plug ignited combustion, and the ignition can be delayed. This ignition delay has a pronounced effect on the production of NO_x; in an engine with precombustion, for example:

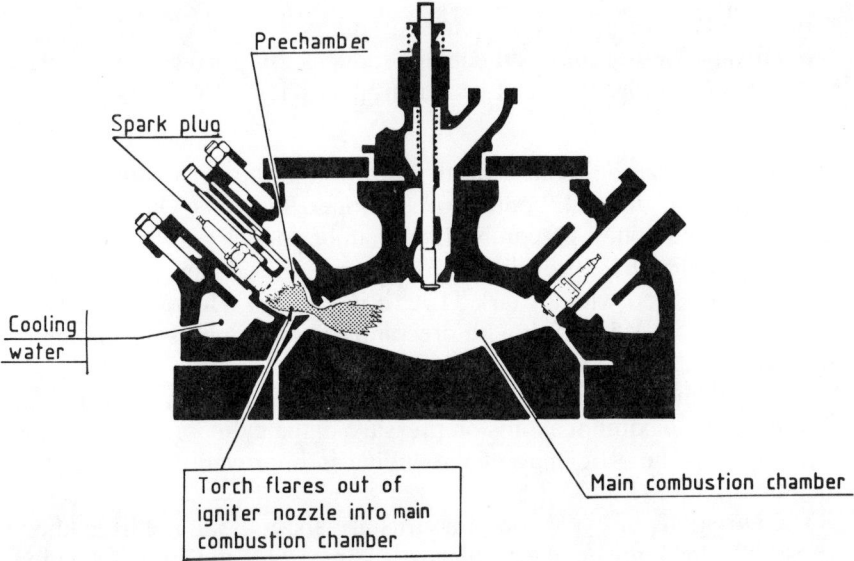

Figure 3-15. Ignition through precombustion chamber forms a recessed spark plug (left). (Courtesy of Cooper Industries.)

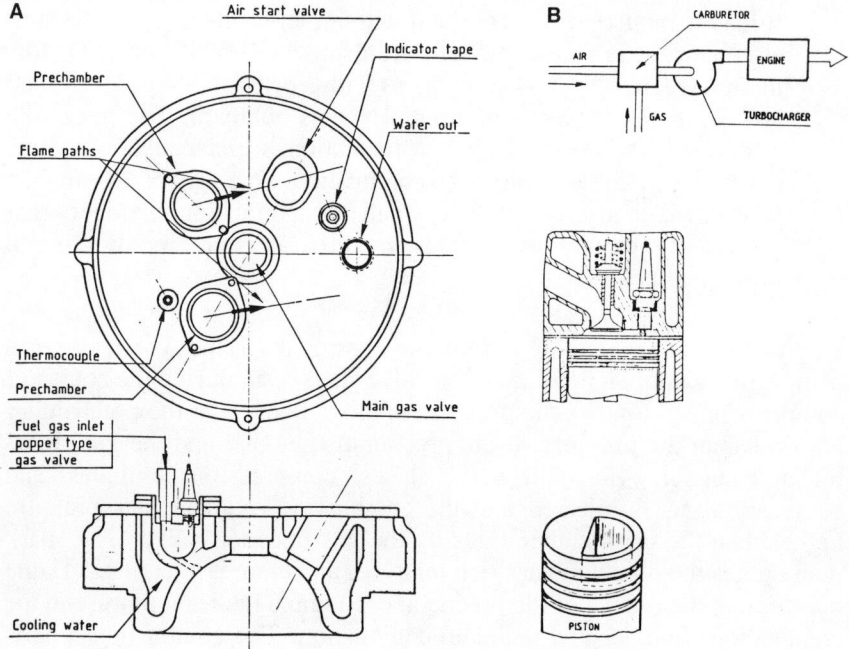

Figure 3-16. (A) Cylinder head design for ignition through two precombustion chambers and low NO$_x$ emissions. (Courtesy of Dresser Clark.) (B) Four-cycle lean combustion system.

- 5° advance before top dead center produces 4.0 g/kWh of NO$_x$.
- 4° advance before top dead center produces 2.7 g/kWh NO$_x$.
- 2° advance before top dead center produces 1.9 g/kWh NO$_x$.

Thus the NO$_x$ can be reduced four-fold through this technique.

For example, the NO$_x$ content of a Cooper Bessemer Quad engine is 15.4 g/kWh without a precombustion chamber and 4 g/kWh with a precombustion chamber (probably 2 g/kWh in a second step that is now in development); and with a Clark TLA, it is 11.4 g/kWh without precombustion and 1.5-4.5 g/kWh with precombustion chambers depending on the load.

This system also has the advantage of creating a much more regular ignition. The maximum explosion pressure in the cylinder remains about constant, and the efficiency of the engine is improved despite the lean fuel-air mixture (Figure 3-8).

The variations of explosion pressures are about 2.5% at full load and 3% at 25% load, the engine regularity is comparable to that of a dual fuel engine. This regularity allows an increase in average explosion pressure and thus mean effective pressure. With a two-cycle engine the explosion pressure of 67 bar for a standard engine can be increased to 82 bar with precombustion chamber engine, and the specific fuel consumption increases less rapidly when the torque decreases; the difference in specific consumption curves at constant torque and constant speed is smaller and allows control of speed and torque at part load at minimized NO_x emissions, without too high a penalty on thermal efficiency.

As a consequence load control for the best NO_x performance would be to vary torque at full speed, since NO_x is highest at maximum torque and lowest at maximum speed at any ambient temperature; NO_x increases 3 to 1 as speed is reduced at full torque from 100% speed to 60% speed.

The prechamber also enables the spark-plug well to be water cooled around the plug support, and this reduces the temperature of the spark plug by 50°–75°C.

The volume of air furnished by the turbo-blower must be increased to obtain a lean fuel-air equivalence ratio of 0.62–0.68. Because the temperature of the exhaust gases powering the turbo-blower decreases as the air-volume increases, the air-volume increase is self-limiting. Also, the air cooler and cooling water system must be sized larger than normal. In general, the air furnished by the turbo-blower for precombustion in a two-cycle engine must be 160%–170% of the stoichiometric, compared to 120%–150% for a normal engine. The cylinder-wall and piston-head cooling is better, and the concentration of burned gases in the air/gas mixture falls from 12% to 8%, reducing the precompression air temperature from 85°C to 75°C and the post-compression temperature from 475°–550°C for a volumetric ratio of 8.8 to 9.6, respectively.

A second method of NO_x reduction is to use four-cycle engines. A richness of 0.71 (air-fuel equivalence ratio of 1.4) is considered to be the flame stability limit for fuel-air mixtures obtained by fuel injection or with a carburetor (four-cycle engines). Nevertheless, it is possible to obtain a richness of 0.60 (air-fuel equivalence ratio of 1.66) with good ignition and combustion if the air-fuel mixture is perfectly homogeneous when ignited and nowhere in the cylinder is the richness lower. This is only possible with a four-cycle engine, which needs no scavenging air, but uses a turbocharger to increase the BMEP to maintain the specific power and to have enough air. The air-fuel mixture is obtained in the carburetor and boosted through the turbocharger, which essentially homogenizes the mixture (Figure 3-16B). The combustion chamber is made compact to increase turbulence and combustion. The chamber has a flat cylinder head and a piston with a deep hollow in front of the spark plug. The ignition requires a minimum voltage of 25 kV at the spark plug and a spark duration of 600 ms, longer than that required by conventional electronic ignition systems. The volumetric ratio can be increased to 12 with natural gas fuel. Then the lean mixture retards the detonation limit, and the thermal efficiency can be maintained as high as 35%. The mean ef-

70 Drivers for Rotating Equipment

fective pressure is 11.5 bar, thanks to the turbocharger. With such lean combustion, it is possible to decrease the NO_x content to 125 ppmv at 3% O_2 at full load (about 0.8 g/kWh).

As reference, diesel engines have an NO_x emission of 8 to 9 g/kWh, much less than gas engines, and the emission level can be reduced to 4–5 g/kWh by an increase in the quantity of combustion air and by retarding the fuel-injection timing.

In addition to the preceding methods, reduction of pollutant emissions can be obtained by adding catalytic converters to both fuel-rich engines (naturally aspirated engines) and fuel-lean engines (turbocharged engines):

1. To use a reduction catalyst with a naturally aspirated engine, the fuel-air ratio must be maintained within narrow limits to have less than 1% oxygen in the exhaust gas at a temperature between 450° and 630°C. The catalytic converter is a "three way" catalyst for control of NO_x, CO, and HC with platinum, rhodium, and palladium. The catalyst should be charged on a metallic or ceramic support in a horizontal plane, with a thickness of 100–130 mm. Through-gas velocity should be 1 m/sec. The maximum effectiveness corresponds to a temperature of about 550°C. The amount of free oxygen present in the exhaust is an especially vital consideration (Figure 3-17).

   The air-fuel equivalent ratio λ, is controlled by a lambda probe (Figure 3-18) in the exhaust gas and is maintained between 0.98 and 1. The lambda probe is the size of a spark plug and is a ceramic

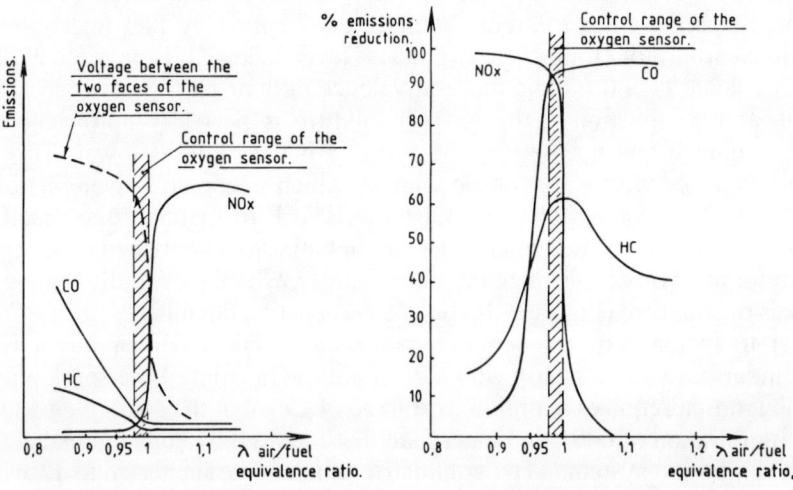

Figure 3-17. Air-fuel ratio influence on pollutants and catalytic conversion.

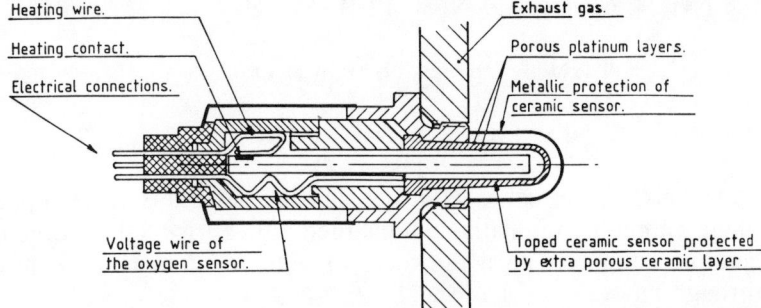

Figure 3-18. Lambda probe.

element charged with zirconium dioxide (Zr O_2) stabilized by yttrium dioxide (YO_2) conductive to the oxygen ions in a large range of temperatures. The element is protected by a slotted metallic tube, set in exhaust gas. The inside of the element communicates with ambient air. The inside and outside surfaces are plated with a thin porous platinum layer working as electrodes; the outside platinum layer is protected with a thin extra porous ceramic layer. When there is a difference in oxygen concentration, there is a voltage difference between the two platinum surfaces. The free oxygen is 0.3–0.5% for an air-fuel ratio equivalence of 1(λ = 1) and combines entirely with carbon monoxide (CO) and hydrocarbons (HC) of the exhaust gas by catalytic conversion on the platinum layer with a resulting abrupt fall in voltage difference from 750 mv to less than 200 mv.

The lambda probe works correctly between 250° and 850°C and is fitted with an electric heating lead to have the right temperature during the starting period of the engine.

The probe becomes worn with the thermal and mechanical stresses and is sensitive to some catalyst poisons, such as lead, zinc, and phosphorus, which come from the fuel or from the lubricating oil. The lambda probe signal controls the air-fuel ratio between 0.98 and 1, with a control loop acting on a bypass on the air or on the fuel gas of the carburetor (Figure 3-19).

2. For supercharged engines the free oxygen content in exhaust gas is high. Since conversion of NO_x requires a reducing environment, oxygen in the exhaust makes it necessary to inject ammonia as a reducing agent prior to the converter. The NH_3 is introduced on a one-to-one basis with NO_x moles, and NO_x is reduced to nitrogen by reaction with ammonia. It involves a catalytic reduction in the flue gas in a controlled temperature reactor containing the catalyst

$$4\ NH_3 + 6\ NO \rightarrow 5\ N_2 + 6\ H_2O$$

72   Drivers for Rotating Equipment

$$8 \text{ NH}_3 + 6 \text{ NO}_2 \rightarrow 7 \text{ N}_2 + 12 \text{ H}_2\text{O}$$

$$4 \text{ NH}_3 + 4 \text{ NO} + \text{O}_2 \rightarrow 4 \text{ N}_2 + 6 \text{ H}_2\text{O}$$

$$4 \text{ NH}_3 + 2 \text{ NO}_2 + \text{O}_2 \rightarrow 3 \text{ N}_2 + 6 \text{ H}_2\text{O}$$

The catalyst is the same as for fuel-rich engines and is made of platinum, rhodium, and palladium charged on a horizontal metallic or ceramic support with a thickness of 100–150 mm and having a through-gas velocity 1 m/sec.

The catalyst efficiency depends on the catalyst constitution and on the exhaust gas temperature (Figure 3-20). With time, catalyst efficiency drops and the temperature window shifts to a higher level. The catalyst effectiveness also depends on the NH$_3$ injection rate (Figure 3-21), thus ammonia controls are essential. Catalyst ef-

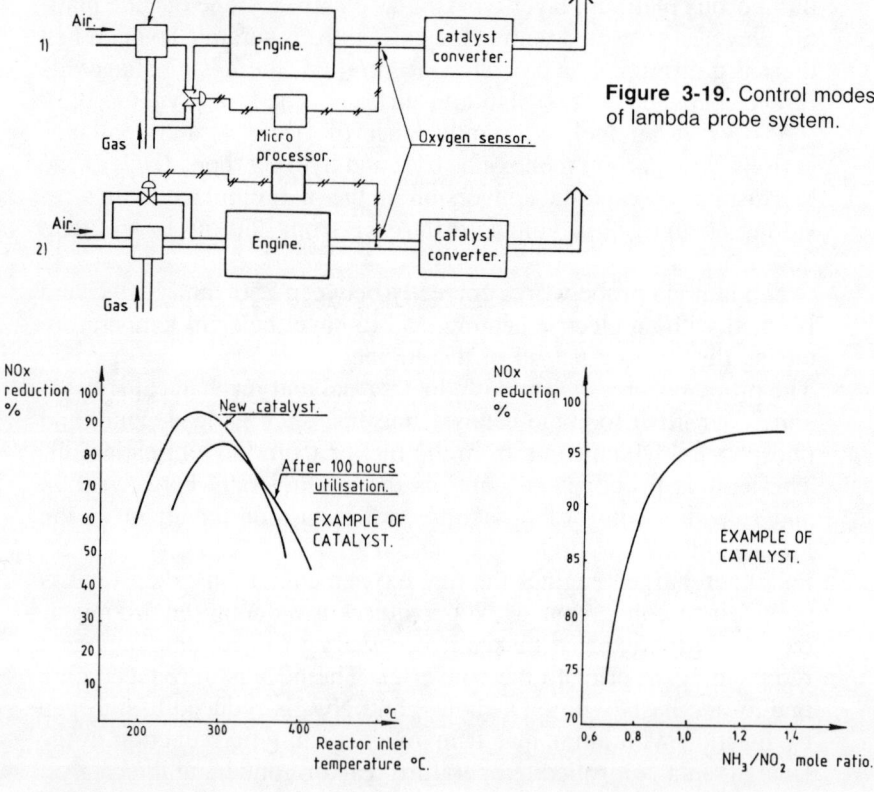

Figure 3-19. Control modes of lambda probe system.

Figure 3-20. Catalyst efficiency vs. exhaust gas temperature.

Figure 3-21. Catalyst efficiency vs. ammonia injection rate.

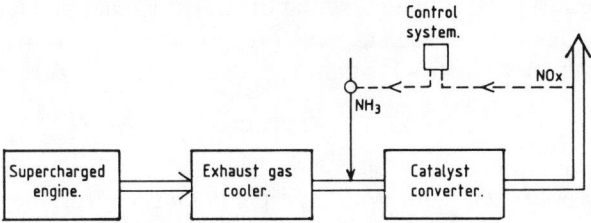

Figure 3-22. Ammonia injection control.

fectiveness drops also when the superficial through-gas velocity increases.

It is possible to reach an NO_x 90% reduction rate in exhaust gas, but the control system (Figure 3-22) is complex, its automation is difficult to achieve since a way to measure NO_x content in exhaust gases is not yet industrially developed. The fuel-rich engines with carburetors are cheaper, simpler, and require less maintenance than supercharged fuel-lean engines but their thermal efficiency is lower.

IGNITION

Gas engines are equipped for spark ignition. Because the electric spark supplies relatively little energy, the air/gas ratio of spark-ignition engines must be carefully controlled. By contrast the injection of gas oil to provoke ignition in dual fuel engines amounts to 4%–7% of the total heat supply, and the dual fuel engines can tolerate wide variations in the air/gas ratio, with consequent good thermal efficiency, wider variations in speed, and more regular explosion pressures.

Contemporary ignition voltage, which must be proportional to the density of the mixture, reaches 40 kV. The conventional ignition system includes, for each cylinder, a low voltage magneto with a distributor and a circuit breaker supplying the high voltage coils. It is of benefit to use multiple ignition points, to reduce the duration of the combustion, and to improve efficiency. Gas engines generally have two spark plugs, but rarely more, because more than two reduces the strength of the cylinder head, limits the cooling capacity, and complicates the ignition system, while increasing costs and maintenance.

Ignition by a battery instead of a magneto is not powerful enough.

Electronic ignition systems offer the advantages of no distributor, no circuit breaker, and consequent greater precision of ignition-point spacing without wear, a low-voltage impulse current with a very sharp front, and a low inductance secondary circuit. The sharp fronted impulse current achieves high voltages with less energy, and the reduction in the in-

74 Drivers for Rotating Equipment

ductance reduces the voltage oscillations, all together reducing wear of the spark plugs and increasing their life (Figure 3-23 A and B).

One of the first electronic ignition systems was the Bosch Pulstronic comprising:

1. A cadmium-nickel battery and a charger.
2. An impulse generator with multiple horseshoe iron magnets driven by the main engine and turning in a stator having a receiving coil per cylinder.
3. An electronic transistorized commutator per cylinder.

This system had the disadvantages of a battery, and spark duration of 1,400 to 2,900 microseconds longer than the spark duration for conventional magneto systems (650 to 1,600 microseconds).

The systems used at present time are self sustained, producing their own energy. One of the most popular is the Bendix S.S ignition system which originally had a secondary voltage of 30 kV, increased now to 40 kV, with a spark duration of 100 microseconds. It includes four parts as follows:

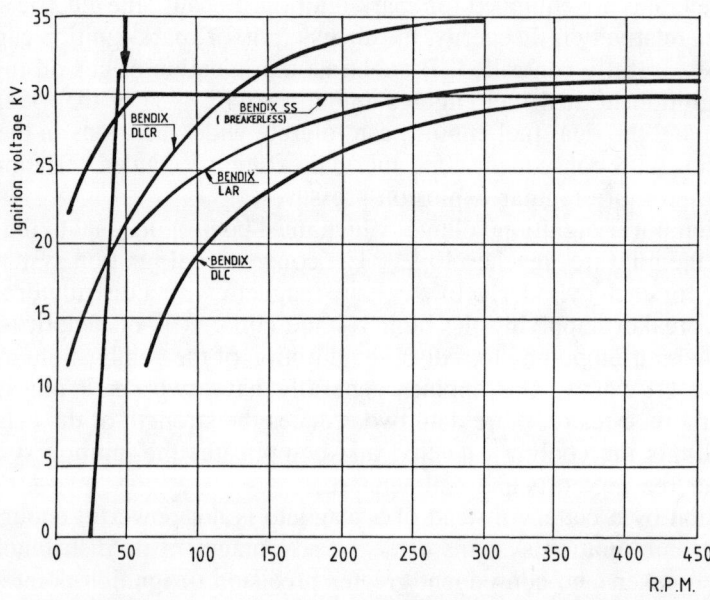

Figure 3-23A. Ignition voltage vs. engine speed for some systems.

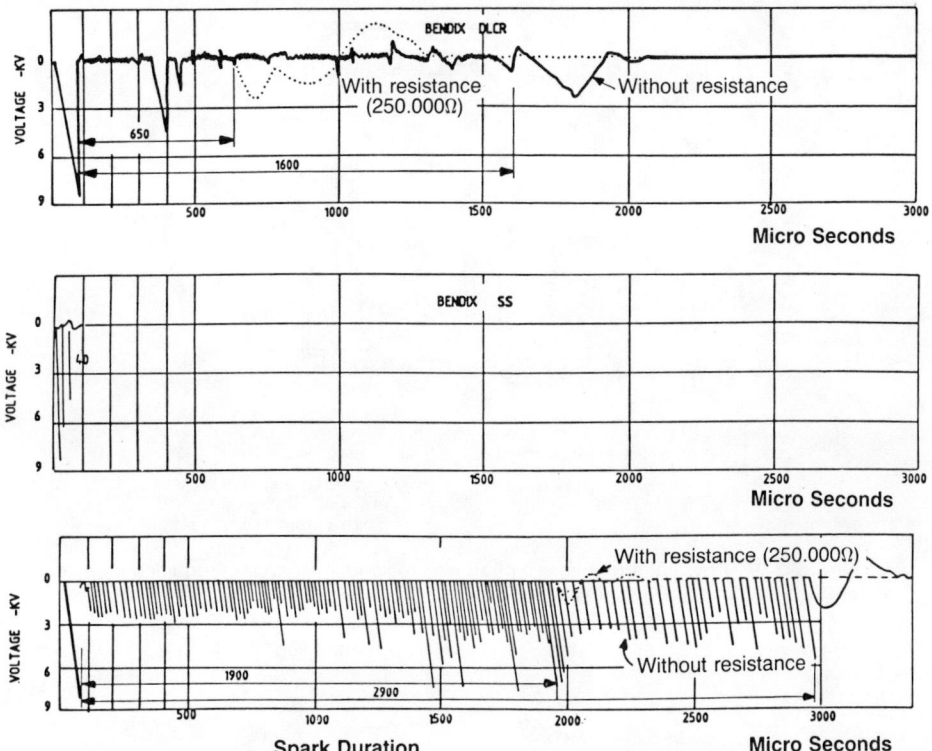

Figure 3-23B. Efficiency comparison of different ignition systems.

1. A double alternator with rotating magnet and three phases (Figure 3-24).
2. Energy storage in the form of condensers charged by the alternator.
3. An electronic distributor turning with the alternator and having a rotating magnet passing fixed coils, one per cylinder. The alternator turns at the same speed as two-cycle engines and half the speed of four-cycle engines.
4. Spark plugs (one or two per cylinder) with incorporated (Figure 3-25) or separate transformer coils mounted near the spark plug to reduce the length of the high voltage leads. The connected leads between the ignition system and the spark plugs thus have a low voltage that suppresses the irregularities in the ignition voltage to the spark plug.

The alternator consists of a 26-pole magnet rotating within a 24-pole, three-phase wound stator-laminator assembly that produces a constant

76 Drivers for Rotating Equipment

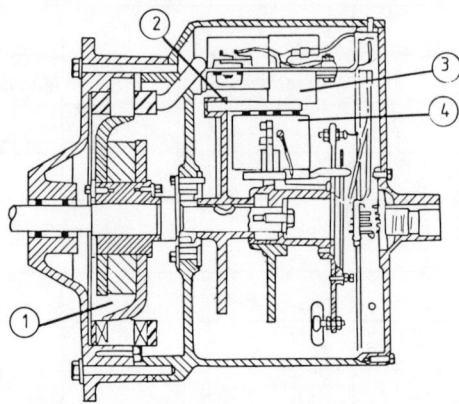

① Two three-phase alternator
② Energy storage by capacitors
③ Rotating magnet arm
④ Cluster of pick up coils to generate small voltage pulses to switch a silicon controlled switch

Figure 3-24. Physical arrangement of an electronic ignition system requiring no external energy source. (Courtesy of Bendix.)

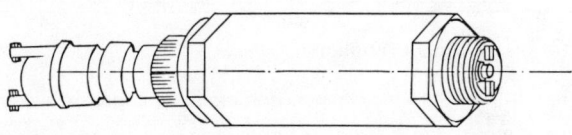

Figure 3-25. Spark-plug and transformer coil assembly. (Courtesy of Bendix.)

flow of AC energy. The AC current produced in the alternator is fed through a full-wave rectifier (D_1–D_4) to provide a pulsation DC current to charge the tank capacitor C_1 (Figure 3-26). The current direction through C_1 from the rectifier is always such that the top of the capacitance is positive, regardless of alternator capacity. Zener diodes Z_1 and Z_2 regulate the upper limit of the voltage applied to C_1 at a nominal 360 volts. This prevents the discharge voltage from exceeding the operating range of the silicon-controlled rectifier (SCR).

The capacitor C_2 is changed to about 20 volts by the voltage divider R_1, R_2; 20 volts is the upper limit voltage of the very sensitive silicon-controlled switch (SCS).

As the drive shaft rotates the vane of the trigger arm attached to it, it passes close to the pulse coil P in the pickup coil, inducing a low voltage

Gas Engines 77

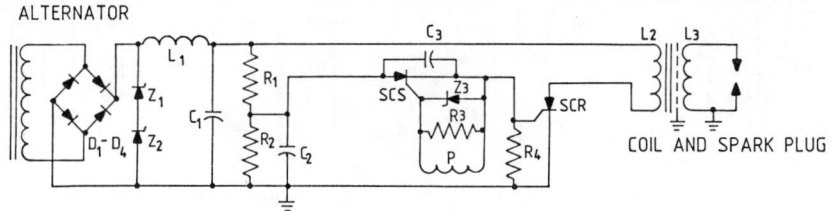

Figure 3-26. Circuitry for an electronic ignition system with no circuit breakers. (Courtesy of Bendix.)

in the coil. The positive portion of the voltage appears in the gate of the SCS and is sufficient to turn on the SCS. This in turn allows the voltage of condenser C_2, which is between the gate and cathode of the silicon rectifier (SCR), to make that transistor conductive so that C_1 can discharge rapidly through L_2. A high voltage of 40 volts is induced in the ignition coil secondary L_3 producing a spark across the air gap in the spark plug.

Coil L_1 provides an inductive "boost" after the spark plug is fired to speed up the recharge of C_1. Zener diodes Z_3 serve to bypass the initial negative pulse produced as the trigger arm vane approaches the pulse coil P and regulates the positive-going voltage as the vane moves away from the coil. Resistor R_3 serves to limit the current through the gate of the SCS. Resistor R_4 serves to limit the SCR current during firing.

Spark Plugs

The spark plugs are of the type with three or four ground electrodes and one central voltage electrode. The electrodes should protrude markedly in order to be swept by the gases and thus kept clear of deposits. The temperature at the end of the electrodes should be between 480°C and 650°C. Above 850°C there is rapid erosion of the electrodes and the phenomenon of auto-ignition. Below 320°C the voltage must be increased for good ignition, and lowering the temperature from 540°C to 320°C requires a voltage increase of 10%. Thus the spark plug should be adapted to the motor and the cylinder head.

The higher the volumetric ratio and combustion temperature, the more the spark plug needs to be cooled. And the required spark-plug cooling varies. Cold spark plugs for gasoline engines and long spark plugs for gas engines are well cooled. Hot spark plugs for gasoline engines and short spark plugs for gas engines are less cooled (Figure 3-27).

Long spark plugs cannot be put into cylinder heads made for short spark plugs, because they would stick out and hit the piston. The spark

78 Drivers for Rotating Equipment

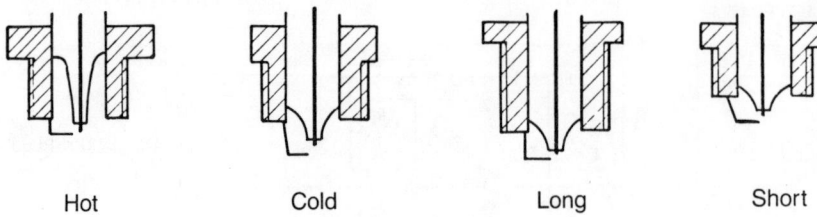

Hot   Cold   Long   Short

Figure 3-27. Spark-plug configurations.

plug gap (spacing of electrodes) is 0.25–0.30 mm. The life of a spark plug is 2,000 hours with magneto ignition, and can reach 8,000 hours with electronic ignition. When their gaskets are made of copper, spark plugs tend to come loose when the operating temperature is higher than the copper's annealing temperature. The gasket expands and is annealed in permanent deformation so that when cooled there is some leakage, which reduces the efficiency and the power of the engine, and leads to corrosion of the spark plug seat. More and more steel gaskets are used in driver engines.

Spark plugs must be screwed in tight, because they are cooled by heat conduction through the threading. It is very important that the exterior insulation of the spark plug, as well as the high-voltage cables, be kept clean. It is especially necessary to be very careful to not put paint on the high-voltage cables, which lose their effectiveness as soon as they touch a ground. It is preferable to apply a negative voltage to spark plugs; a positive voltage must be 20%–25% higher.

Explosion-Proof Ignition

The voltage connections between the impulse generator and distributor and the individual spark plugs are normally as follows:

1. For low-voltage connections two wired cables are inside an insulated or braided metal sheath with either multiple pin plugs plugged at the ends or a threaded muff fixed at one end to the distributor and at the other end to a junction box (Figure 3-28). From the junction box to the coils, two wired cables in a conduit run along the walkway at the level of the top of the cylinders with a connection to each coil inside a sheath in braided metal plugged and fixed with a threaded muff to the coil (Figure 3-29).
2. For high-voltage cables between coils and spark-plugs with push-on sleeve connections at each spark plug a copper rod extension is used

Gas Engines 79

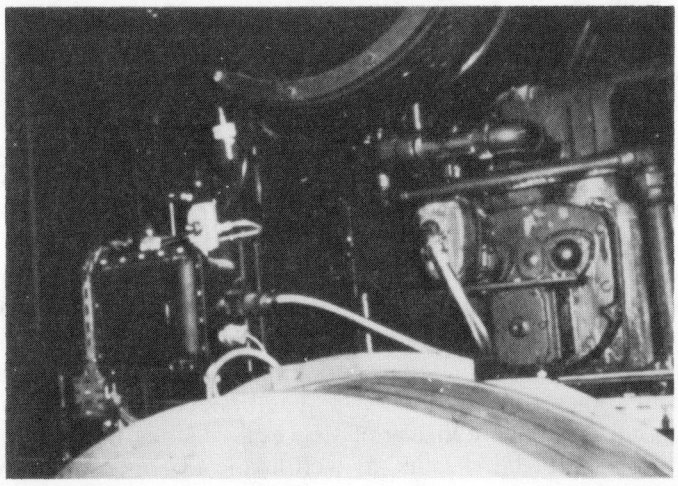

Figure 3-28. Cables under braided metal sheath connected between the distributor and impulse generator on a junction box.

Figure 3-29. Shielded low voltage cable between coil and distribution tube, as well as the coil on the cylinder head and a copper connecting rod in the spark-plug well.

80 Drivers for Rotating Equipment

when the spark plug is at the bottom of a well. This avoids wear and heat damage that would occur to a flexible sheath against the well wall.

The cable is of stranded copper wire (tinned or not), with insulation resisting up to 85°C. The insulation can be of polyvinyl chloride or polyethylene or synthetic rubber. The outside of the coil is connected to the ground (Figures 3-29 and 3-30).

When an explosion-proof ignition is required, all the cables, including the high-voltage cables, are shielded in a braided metal sheath; the metal sheath for the high-voltage cable is connected to the coil by a threaded metal sleeve on one end and to a spark plug shielded inside a metal envelope on the other end (Figure 3-31). The braided metal sheaths are all connected to grounds by means of sleeves.

These braided metal sheaths, as well as the sleeves of shielded leads and spark plugs all function as capacitors and the capacitance increases with the length of the lead. The voltage available to the spark plug is reduced proportionally as this capacitance increases (Figure 3-32). This results in a weaker ignition. Also, if the spark plug shows the slightest leak

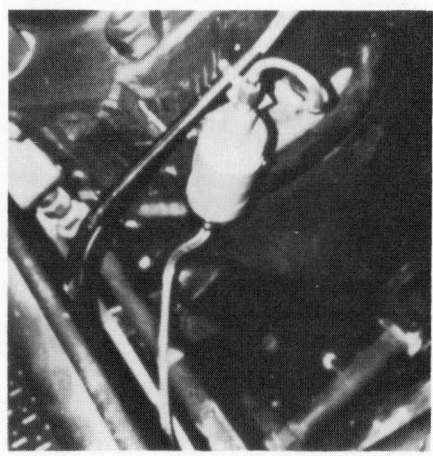

Figure 3-30. Shielded ignition system, explosion proof coil, and shielded high voltage cable (braided conduct or wire).

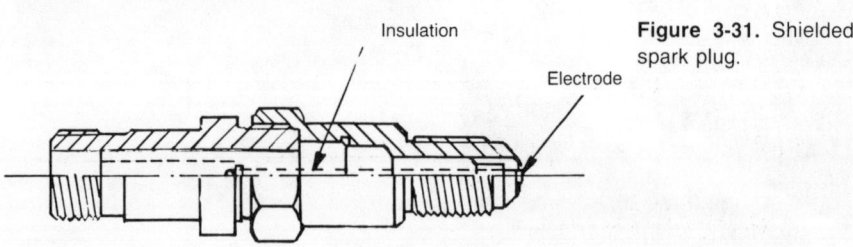

Figure 3-31. Shielded spark plug.

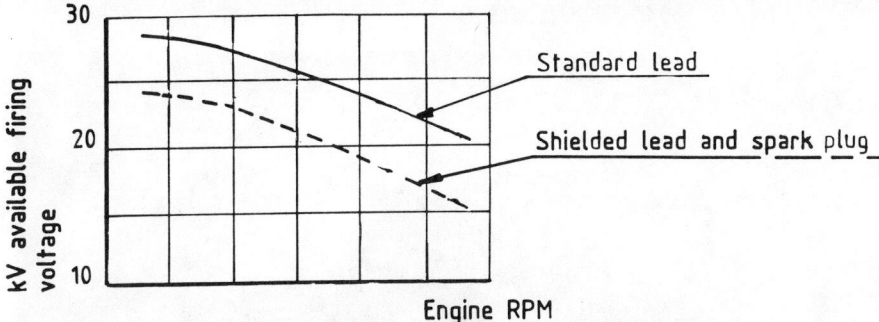

Figure 3-32. Effects of shields on available firing voltage. (Courtesy of Ross E. Nielsen.)

in its insulation, air enters through the sheath, ionizes between the wire and sheath, and creates electrical leaks; the temperature goes up inside the sheath, and the spark plug connection deteriorates. Therefore, there is a risk of misfiring with explosion-proof ignitions, due both to the spark plug and to the sheath. Consequently it is necessary to:

- Keep the high-voltage lead as short as possible by placing the coil directly on the cylinder.
- Frequently (about every 600 hours) change the high-voltage wires and the spark plugs in order to have good insulation and good contacts.
- Continuously check the state of the spark plugs for any sign of leaks around the insulation; any trace of gas leak at the spark plug or any indication of poor contact mandates immediate replacement of the high-voltage wire and the corresponding spark plug.

In sum, the ignition system must be maintained perfectly and regulated. It is much preferred to have a nonexplosion-proof ignition and one or more dangerous-atmosphere detectors located above the cylinders. It should be pointed out that the risk of igniting a mixture of the gas above the cylinders is very small if not negligible. In some cases it is required that the coils even be enclosed in explosion-proof boxes (Figure 3-33).

One method of avoiding the metal sheathing on high-voltage cables, as well as shielded spark plugs, has been used successfully on the engines installed in the Ruhrgas compressor stations. The spark plugs there are conventional; the high-voltage cable has double insulation of PVC sheathing around a conventional cable, with one end soldered onto the head of the spark plug and protected by a copper sleeve so as not to break the soldering; and all is enclosed in an insulating sleeve (Figures 3-34 and 3-35). This type of connection is accepted as explosion proof by the German safety service.

82 Drivers for Rotating Equipment

Figure 3-33. Explosion proof boxes and shielded high voltage cable installed on a gas engine.

Figure 3-34. Shielded high voltage cable.

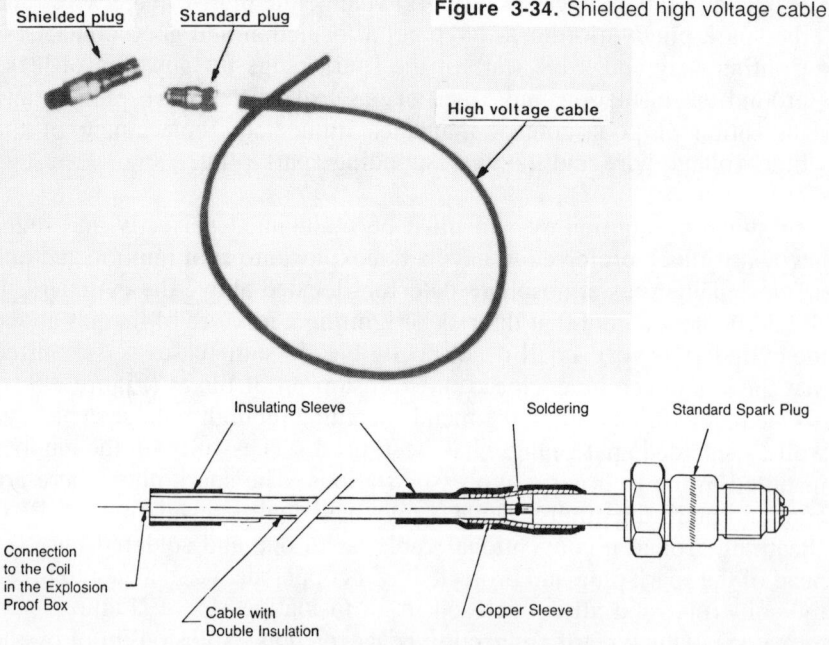

Figure 3-35. Shielded high voltage cable.

Gas Engines 83

Reducing Interference in Radio Transmission

Engine ignition systems can cause static for nearby radio and television receivers; and it is sometimes necessary to eliminate such static. The disturbance field at 30 m from the engine should not go over 50 µv/m for frequencies lower than 75 mHz (megahertz) and 50–120 µv/m for frequencies between 75 and 250 mHz. The effect increases linearly as a function of the frequency.

Metal braided shielding is generally not enough, but it is not necessary to go so far as shielded spark plugs with their associated ignition problems. Rather it is enough to have one end of the shielding fixed to the coil, as with a shielded ignition, and the other end grounded at the cable's entrance to the spark plug well (Figure 3-36).

Protection against static requires a helical resistance of 5,000–20,000 ohms on the high-voltage wires between the coil and the spark plug. This resistance also has a self-inductance. This protection can be obtained by:

- Resistant high-voltage cables with a core of nylon wires, a ferromagnetic tube of a ferrite composition and a binder, and a shield of a helical wire (Figure 3-37). The resistance of such cables varies with the

Figure 3-36. Ignition cable with grounded shield.

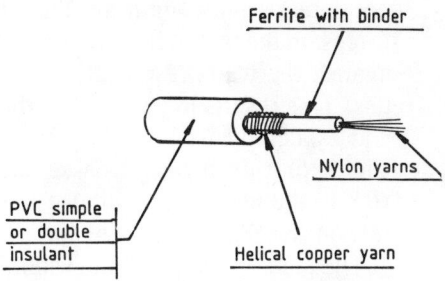

Figure 3-37. Construction of no-static ignition cable.

length and manufacture. There are two resistances, 1,400 ohms/m for fixed engines and 5,600 ohms/m for automobile engines where the reduction in interference should be the greatest.
- High-voltage resistant cables with a conductor of carbon fibers impregnated with insulation to give a resistance of 18,000 ohms/m.
- Normal high-voltage cables with a stranded copper core with resistances in series (Figure 3-38), as on automobile engines, where there is a resistance of 1,000–1,200 ohms in the insulating connection between cable and spark plug, and of 8,000–10,000 ohms in either the distributor's rotating arm or the connection to the distributor of the cable between the coil and distributor.

The reduction in interference is improved by a low capacity condenser of 2–5 μF on the low voltage outlet of the coil.

Ignition Quality Control

The quality of ignition is of prime importance. Any irregularity that causes a poor or no ignition of a cylinder causes that load to be put onto the other cylinders, because the regulator increases the amount of fuel mixture needed to maintain the speed, and the torque control system does not detect a poor ignition. The explosion pressure and consequently the forces on the piston head and piston rings, as well as the force of the rings against the walls (especially the front ring), all surpass the normal values. This results in ruptures in the lube oil film, wear of the piston rings, and especially in two-cycle engines, wear of the inlet and outlet ports. The explosion pressure passes through the resistance of the lead piston ring to the next ring, and also the flame, which leads to a high temperature at the bottom skirt of the piston and consequent seizing.

The state of the ignition system should therefore be frequently verified and automatically controlled by an instrument that measures the temperatures of the exhaust of each of the cylinders and compares them to the average of all the cylinders and that stops the engine as soon as one cylin-

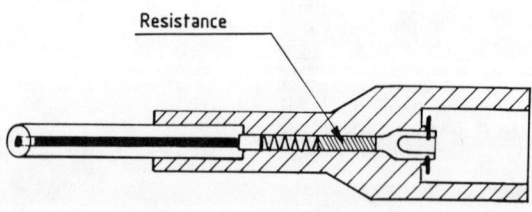

Figure 3-38. High voltage cable-plug connection with 1,100 ohms resistance in series.

der no longer ignites correctly. Temperature detectors with thermocouples are also used in the cylinder heads to measure the wall temperature of the explosion chamber. These instruments particularly detect detonation and pre-ignition.

PISTON RINGS

The clearance between the piston and cylinder walls is sealed by means of three or four cast steel piston rings. These rings press against the cylinder wall because of their elasticity when at rest; during operation the pressure in the cylinder forces them laterally against the cylinder wall, as well as vertically against the edges of the grooves where they are seated—the more so, as the cylinder pressure gets higher.

The rings are discontinuous, with a gap through which the gas pressure can partly pass the lead ring and pass on to the back-up rings. This gap must be enough for the force of the explosions to be distributed over the rear as well as the front rings, but it should not be so much as to allow leakage or "blow-by." The gap can be straight, slanted, or in steps with an overlap. Slanted gaps reduce blow-by.

There should always be an oil film between the piston ring and the cylinder wall. If not, the rings heat up, sometimes melting, seizing, and stripping off metal. A chrome-plated outside surface reduces the risk of seizing, because the two dissimilar metals with soldering or stripping will not melt at once.

Piston rings can generally be one of three types:

1. Wide rings with the wear face perpendicular to the groove seat (Figure 3-39A). This type consumes more power and requires more oil because the friction is greater.
2. Narrow rings with the wear face sloped 1°-2° from the line of the piston (Figure 3-39B). These rings have a shorter break-in period for contact to be established around the whole circumference by wearing off high spots. The oil consumption is lower; these rings are less liable to seizing; and the lives of the rings and cylinder wall are longer.
3. Narrow rings tilted in their groove-seats and with slanted wear faces (Figure 3-39C) so that the amount of bending they experience at rest is increased by the pressure during operation. The seal at the seat of these rings is improved. This type of ring is the best.

A slanted-face ring favors formation of oil wedge in advance of the seal on the upward stroke, and scrapes the oil on the downward stroke. Because the oil film is kept thin, heat transfer is better, and the piston and rings are better cooled. Also, the contact surface is smaller.

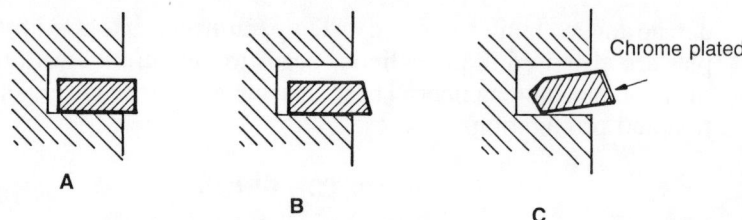

Figure 3-39. Piston rings in their grooves.

The piston has on its skirt away from its head one or two oil-wiping rings of cast steel. These have between the ring and the seat a spring (Figure 3-40) that forces the ring against the cylinder wall, because the rings are not exposed to the cylinder pressure and are pierced with holes to allow oil to flow by. These oil-wiping rings serve to control the oil consumption and maintain the correct level, and to distribute the oil uniformly over the cylinder wall.

The groove-seats holding the rings on the piston must be perpendicular to the piston wall; if they are not they must be retooled. The rings should be tested by sliding them into the cylinder without the piston to assure that the clearance, as determined by light leakage, is exactly the same at all points of the periphery. If not, the ring touches the cylinder wall only in places and there is risk of seizing, at least during break-in before wear has made the clearance uniform. At the same time, the walls of the cylinders must be perfectly circular and straight—no ovals nor cones beyond the wear tolerance—otherwise they must be honed with a wetstone in a rotating movement that creates slight transverse striations to retain oil during break-in.

SUPERCHARGING: INFLUENCE OF ALTITUDE AND SUPERCHARGING AS A FUNCTION OF AMBIENT TEMPERATURE

The horsepower of an engine is proportional to

1. The weight of the air-fuel mixture admitted per cycle.
2. The number of cycles per second.

Supercharging affords an increase in the weight of mixture admitted per cycle. It can be accomplished with a compressor (generally Roots type) driven by the engine, or by a separate turbocharger. Since 1930, diesel engine turbochargers have been powered by exhaust gases.

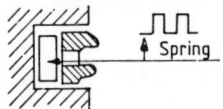

Figure 3-40. Oil wiping ring.

Supercharging increases the specific horsepower of the engine, which is to say the power of a given cylinder volume, and which is proportional to the density of the fuel-air mixture at the end of its admission to the cylinder. Also, supercharging increases thermal efficiency through improved recovery of the energy in the exhaust gases, as well as improved combustion efficiency. Furthermore supercharging reduces oil consumption and wear on the pistons, liner, and rings, because a continually positive pressure on the piston causes the rings to keep the same position in their grooves, and the reduced wear makes the engine tend to pump less oil.

Supercharging pressures are currently on the order of 300–500 mbar but can go as high as 3 bar. This increase in pressure, with its correspondingly higher cylinder pressures and temperatures during the power cycle, creates two sorts of problems, mechanical and thermal. The mechanical problems are mostly due to increased friction between the rings and the cylinder liners, and to a lesser extent to increased maintenance of the axle and bearings. The more important thermal problems are due to a tendency for detonation because of the higher intake temperature and to the difficulty of cooling. In current engines exhaust gas temperatures are limited to 650°C; and beyond that there is risk of burning the upper face of the piston and forming carbon deposits in the piston-ring grooves.

This last problem is partly solved by increasing the excess air, which reduces the temperature at the end of combustion, by cooling the air after it leaves the turbocharger, and by well scavenging the cylinder before compression in order to cool the walls. (Such scavenging is possible only with engines where the fuel is injected to the cylinder and not mixed with the air before injection.)

Supercharging with a turbocharger makes possible recovery of only a part of the heat in the exhaust, because the adiabatic efficiencies of the turbine and compressor must be taken into account, and these will already be as high as possible. Also, the turbine exerts a back pressure on the exhaust; and that back pressure increases as the supercharging pressure is increased, to reduce the expansion of the combustion gases and thus the horsepower of the engine.

At a sacrifice of some of the engine's power the maximum energy in the exhaust gases can be recovered by means of an expansion turbine (Figure 3-41). This is the case of the free-piston generator described in Chapter 4.

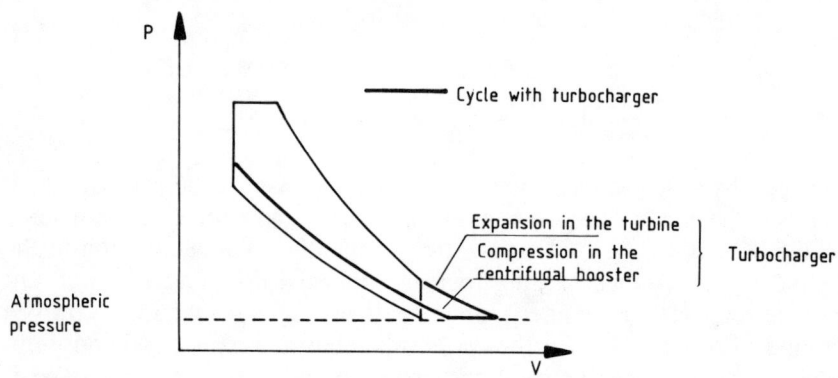

Figure 3-41. Engine cycles with and without turbocharger and exhaust turbine.

Supercharging Systems

Four-Cycle Engines

The most common system consists of a turbocharger with an expansion turbine on the exhaust gases and a radial-type centrifugal compressor with one impeller driven by the turbine. Air aspirated across a filter is compressed and then usually cooled before admission to the cylinders.

The centrifugal supercharging compressor can operate at low pressure (300–500 mbar) and have a large flow, of which some 60% is used for scavenging and cooling the cylinders by means of simultaneously opened intake and discharge valves at the exhaust top dead center. The mean effective pressure can be from 8.5 to 10 bar. The discharge pressure of the compressor can be increased with the load because of the higher exhaust gas temperature and flow. Thus the correct air-fuel gas ratio can be maintained for all loads; and the necessary regulation can be done with butterfly valves upstream of the inlet main, either manually or automatically. The amount of fuel gas injected can be regulated as a function of the load by varying the gas pressure in the inlet collector.

In other instances the centrifugal supercharging compressor can discharge at high pressures, with current values on the order of 3 bar. In this case the engine's intake valve closes before the bottom dead center, and the supercharged air expands to a pressure near 400 mbar for a temperature near 10°C between the moment of the intake valve's closing and the bottom dead center. Subsequently the mean effective pressure reaches 11–12 bar, and the overall thermal efficiency goes over 40% for volumetric ratios on the order of 12. Controlling the intake valve's closing point with the load provides a precise regulation of the air/gas ratio, a correct ignition, and good operating stability. The pressure of the fuel gas remains constant, but the amount of gas injected is varied as a function of the load by varying the length of the injection period (Nordberg "Supairthermal" engine).

Another supercharging device called turbo-cooling is used by Cooper Bessemer (Figure 3-42). In this system the combustion air is compressed twice, first with energy from the exhaust gases and secondly with part of its own pressure energy, before being passed through a cooler and its expansion engine to the combustion cylinders. This procedure affords air intake temperatures below the ambient, as well as effective average pressures of 11–12 bar.

Two-Cycle Engines

Turbocharging two-cycle engines requires more care than for four-cycle engines for three reasons:

- The energy available in the exhaust gases is less because of lower temperatures.
- The scavenging air flows are higher (130%–200%) than for four-cycle engines.
- Startup requires auxiliary blowers to supply the air comparable to that supplied by one piston stroke in the four-cycle engine.

Consequently turbochargers with efficiencies higher than 60% overall are necessary to make turbocharging two-cycle engines more economical than the conventional mechanically driven piston or centrifugal scavenging-air blowers. Even then, the turbochargers are usually powerful

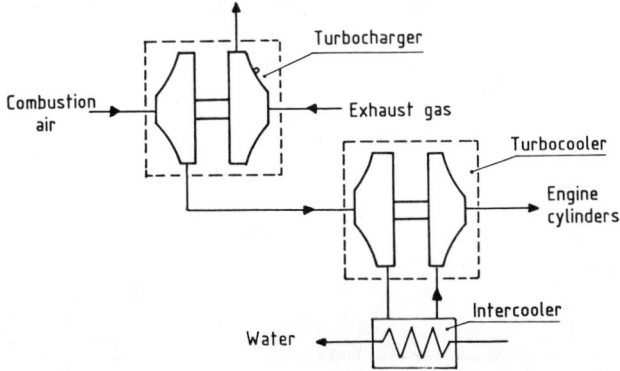

Figure 3-42. Flows of air and exhaust gas in turbocharged and turbocooled systems.

enough only during operation above 70% of the load; and it is necessary to include some system to provide normal scavenging air below this value. Contemporary systems comprise essentially:

- A mechanically driven turbocharger with a hydraulic coupling in an automatic clutch, to maintain the turbocharger at below 75% of speed.
- Means for starting the turbocharger and maintaining its speed, such as, for example, the injection of air compressed in the turbine.
- Series flow through a compressor driven by an expansion turbine on the exhaust gases and a mechanically driven compressor as shown in Figure 3-43.

Supercharging pressure is limited by the pressure existing in the exhaust manifold. The temperature of the air after compression and cooling is on the order of 50°–55°C at the intake to the cylinders.

Because of the higher thermal load of two-cycle engines, relative to four-cycle ones, the mean effective pressures can be only 8–9 bar in order to avoid detonation.

However, supercharging by turbochargers affords a significant reduction in the specific consumption of two-cycle engines through better control of the air/gas mixture, recovery of the exhaust heat, and elimination of scavenging blowers driven by the engine. The air/gas ratio in a two-cycle engine is rather markedly affected by variations in load or speed, and the automatic control of this ratio as a function of the load and speed is more important than for four-cycle engines.

High Altitude Engines

Supercharging at high pressure can maintain the effective average pressure and thus the horsepower of an engine at high altitude, whereas

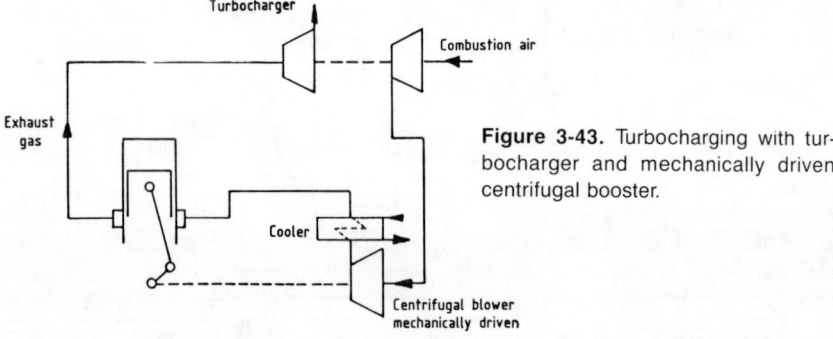

Figure 3-43. Turbocharging with turbocharger and mechanically driven centrifugal booster.

the power of nonsupercharged engines decreases with an increase in altitude (Figure 3-44).

Supercharging as a Function of the Ambient Temperature

In order to improve the competitiveness of gas engines versus gas turbines, whose available power increases as the ambient temperature decreases, American engine manufacturers have offered improved cooling circuits to permit supercharging when the ambient temperature is lower. (This is called ambient uprating.) The nominal BHP of engines is usually given at 80°F (27°C); and it is possible with some motors that are mechanically suited to increase that nominal power by 16% at a temperature of 40°F (4.4°C), the increase in power being linear with the decrease in temperature (Figure 3-45).

This assumes that the temperatures and pressures of combustion and scavenging air, as well as of the oil and cooling water, are maintained as a function of the load (Figure 3-46), which is possible with an appropriate cooling system control that takes account of the increased cooling capacity made possible by a lower temperature. The fan of the water cooler is maintained at maximum speed so that the water temperature is as low as possible. The temperature of the combustion air follows the water temperature; and the pair are regulated as a function of the combustion air temperature. Normally the engine's cooling water temperature should be 6–9°C above the oil temperature (6°C at the inlet of a two-cycle engine,

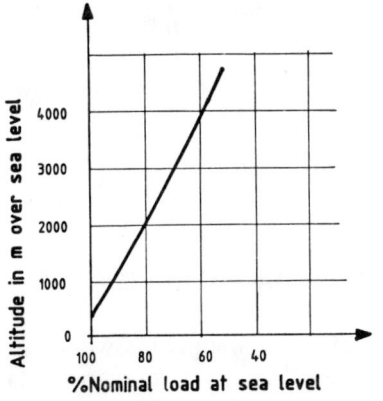

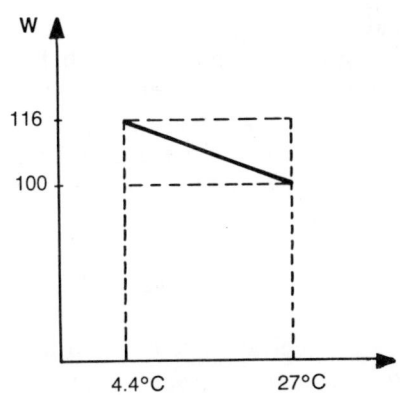

Figure 3-44. Influence of altitude on the power available from naturally aspirated engines.

Figure 3-45. Ambient uprating: power increase with an ambient temperature decrease.

92 Drivers for Rotating Equipment

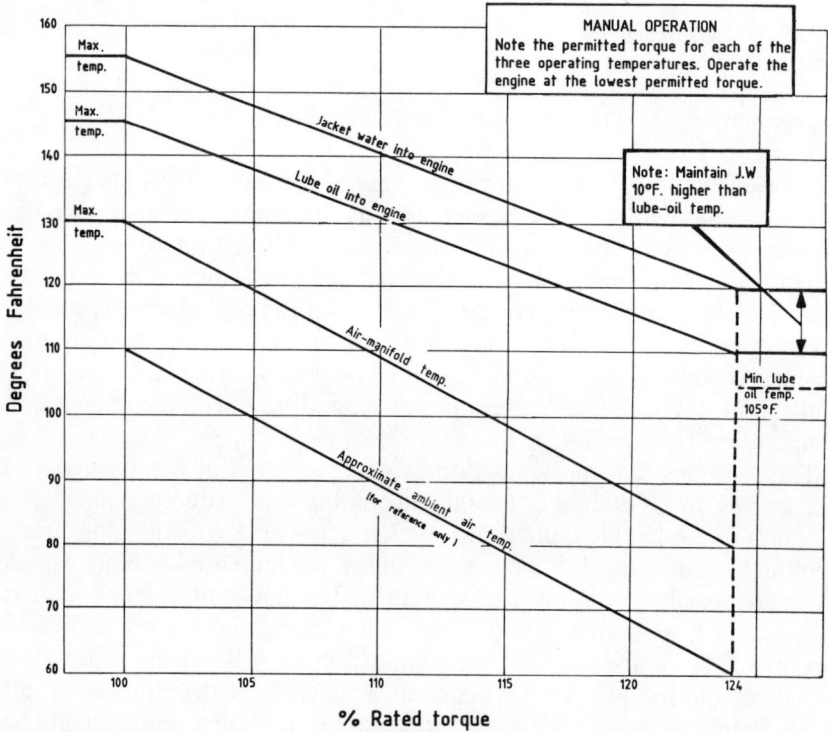

Figure 3-46. Temperature limits at different torques for Clark TCV engines. (Courtesy of Dresser Clark.)

and 9°C at the outlet for a four-cycle engine). The temperature of the combustion air must not go below 16°C.

For a two-cycle engine, the water and oil temperatures decrease with decreasing load. For a four-cycle engine, the temperatures of water and oil at the outlet of the engine are kept constant at 80°C and 79°C, respectively. The oil temperature must not go below 45°C to avoid the formation of sludge.

WEIGHTS, FLOOR SPACE, AND ROTATING SPEEDS

The specific horsepower of an engine, i.e., its nominal power related to its cylinder capacity, is proportional to its speed of rotation for any given mean effective pressure. Therefore an increase in the speed of rotation leads to an increase in the specific power, and to a reduction in weight and required space, if it is further allowed that weight and required space are proportional to the cylinder capacity. However, improvements based on higher speeds involve five additional factors, as follows:

1. Because metal fatigue is proportional to the square of the average speed of displacement, the metallurgical and mechanical quality of the metals can be limiting.
2. The quality of the tooling can be limiting.
3. The quality of the lubrication can be limiting.
4. Because the displacement speed is not so much the direct cause of wear as is a high temperature in the rubbing parts, the quality of the cooling of piston and cylinder liners can be limiting.
5. Because the fluid-flow losses at the inlet and exhaust increase with speed, and because it is necessary to obtain a higher maximum pressure in order to maintain the mean effective pressure at high speeds, the reduction in efficiency can be limiting.

All of these factors explain why manufacturers currently limit themselves to piston-liner speeds on the order of 5 m/s for normal operation and 7 m/s for engines in generator sets, and why they prefer to increase the specific power through improving the mean effective pressure. The maximum rotating speeds are 800 rpm for two-cycle and 1,500–1,600 rpm for four-cycle engines, corresponding to powers less than 300 kW for engines destined for generator sets. Beyond 750 kW, the most common speeds are on the order of 300 to 500 rpm. Great progress has been made in the efficiency and ruggedness of high-speed machines, in order to maintain a place for gas engines in the market for generator sets; and a high speed of rotation is completely compatible with excellent longevity.

Engine Types Currently Available

The available engine types can be classified as average or low power engines, high-power engines, and integrated gas engines with compressors.

The average and low power engines are of 5 to 750 kW, rotating at 600 to 1,800 rpm, with the highest speeds corresponding to the lowest powers. Their efficiency varies from 28% to 35%. These engines are well suited to generator sets (Figure 3-47).

The high-power gas engines are destined for driving centrifugal pipeline compressors, or pumps for oil pipe lines. Their powers range from 750 to 6,000 kW. They have 10 to 16 cylinders in line or in a V; and their rotating speeds range 300 to 500 rpm. They have an efficiency of 28% to 32% without supercharging and 37% to 42% with supercharging (Figure 3-48).

94 Drivers for Rotating Equipment

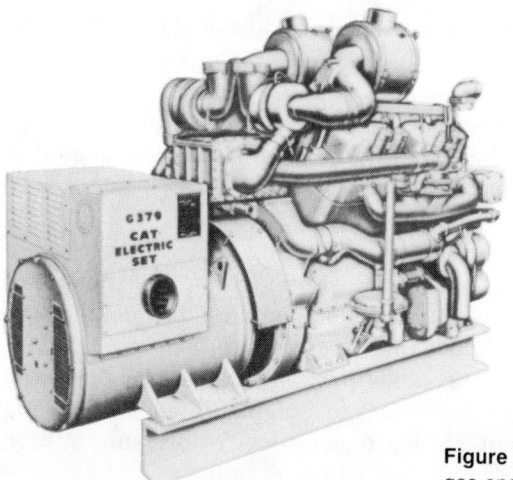

Figure 3-47. Packaged generator set with gas engine. (Courtesy of Caterpillar.)

Figure 3-48. Centrifugal compressor driven by a 2,900 kW engine in a Transcontinental pipeline compressor station. (Courtesy of Nordberg.)

When gas engines are destined for driving reciprocating compressors, they are most often of the integrated type; that is, the frame and crankshaft of the motor and compressor are in common, with the compressor being horizontal (Figure 3-49A). The range of power of engines of this type is 350–9,000 kW, for rotating speeds of 300–475 rpm. Their efficiency is 28% to 32% without supercharging; 35%–40% for four-cycle engines with supercharging, and 32%–38% for two-cycle engines with supercharging. These efficiencies, which are slightly less than those of the independent gas engines already mentioned, are explained by the increase in mechanical losses due to the increase in number of bearings and to assigning the mechanical losses of the crossheads and connecting rod bearings of the compressor to the engine. In powers ranging 100 to 1,000 kW the integrated engine-compressor groups can be mounted on skids with all the necessary auxiliaries, so as to form a completely autonomous package of engine and compressor (Figure 3-49B).

Figure 3-49. (A) Packaged 1,500 kW reciprocating compressor with integral engine having eight cylinders in line; (B) 330 kW integral reciprocating compressor driven by a two-cycle engine. (Courtesy of Dresser Clark.)

GAS ENGINE OPERATION

Operating a gas engine calls for attention to, and control of, the speed, energy consumption, oil consumption, ambient air temperature, ignition, air/gas ratio, fuel quality, mean effective cylinder pressure, and routine maintenance.

Speed Regulation

The limits to the range of speeds of a gas engine are set by the ability of its fly-wheel to control the cycles and the existence of critical speeds at which the crankshaft enters into dangerous torsional vibrations.

Some gas engines are equipped with a fly-wheel to regulate the cycles and dampen torsional vibrations. This wheel is calculated as a function of the minimum required speed. It is a hydraulic dampener comprising a balance wheel floating in a liquid containing silicones inside a casing carefully filled and rigorously sealed air-tight, once and for all, at the factory (Figure 3-50).

The period and amplitude of a torsional vibration depend on the length of the crankshaft, which should be as short as possible. This is the basis on which the two-cycle engine must sometimes be preferred over the longer-shafted four-cycle engine; it also explains why an independent four-cycle engine is less fragile than an integrated four-cycle engine.

Independent gas engines have the ability to operate at speeds varying from 50% to 110% of the nominal, and on demand such engines can even reach speeds as low as 20% of nominal. Integrated moto-compressor groups, by contrast, have a potential speed variation of 60%–100% of nominal; and some engines integrated with compressors have a speed variation limited to 80%–100% of nominal. For all these engines, the critical speed is usually (but not always) above the maximum speed.

Engines can be equipped with vibration detectors, so that crankshaft breaking need not be feared.

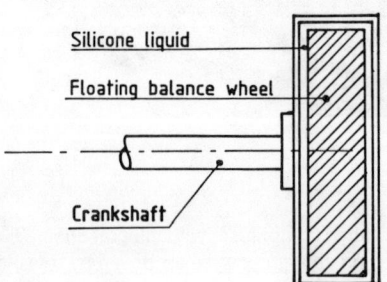

Figure 3-50. Hydraulic dampener.

Energy Consumption

The energy consumption at variable speeds and a constant torque equal to the maximum should be distinguished from the consumption at constant nominal speed and variable torques. Figure 3-51 shows the two curves of energy consumption, one at constant speed and one at constant torque, as a function of the horsepower demanded of two-cycle and four-cycle engines. The constant torque curve is almost horizontal for a four-cycle engine and slightly increasing for a two-cycle, whereas the constant speed curve rises abruptly. This is why it is better to operate a moto-compressor group at maximum constant torque. For comparison, Figure 3-51 includes curves 1 and 2 for a second generation gas turbine (Allison 570 k of 4,800 kW with a thermal efficiency ISO related to the LHV of 30%). Curve 1 is at variable speed, when energy consumption varies as the cube root of the power (also the case of a centrifugal compressor). Curve 2 is at constant nominal speed.

For a fuel-lean engine with fuel-rich precombustion chamber which maintains the combustion, the specific fuel consumption curve at constant speed is much closer to the curve at constant torque. It is because the explosion pressures reduced proportionally to the torque have a much better regularity due to less misfiring. That is the case for engines designed for NO_x reduction. For such engines the load control varies torque at full speed for best NO_x performance.

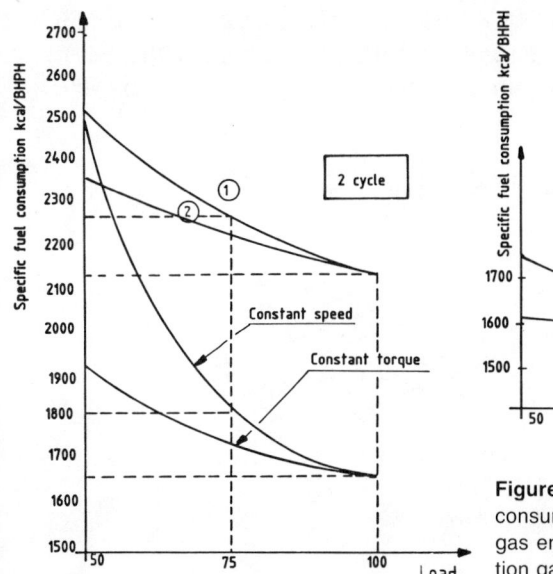
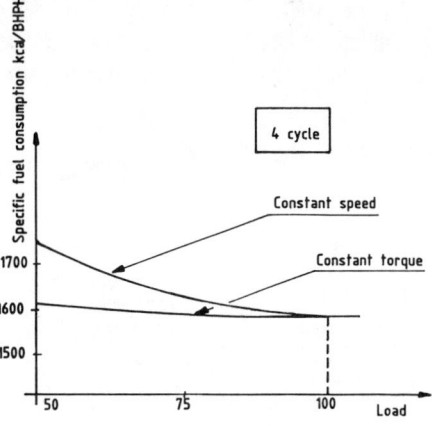

Figure 3-51. A comparison of specific fuel consumptions for two-cycle and four-cycle gas engines, with those of second-generation gas turbines.

Oil Consumption

The oil consumption, including oil changes, can be assumed as 0.95 liters per 1,000 kWh for a two-cycle engine and 0.7 liters per 1,000 kWh for a four-cycle engine. Oil consumption is a function of the quality of the cooling.

Effect of Ambient Conditions on Rated Load

The power of an engine decreases at altitudes greater than 450 m above sea level; but up to 3,000 m the effect of the altitude can be compensated for by using the supercharging air blower. On the other hand, too high an ambient air temperature lowers the available power, according to Figure 3-52. This figure gives the coefficients of available load as a function of temperatures of ambient air and water, which is usually 8°C warmer than the ambient air temperature when the water is cooled in an air cooler.

Ignition Control

As described earlier, the control of the ignition is a function of the speed of rotation, the torque as expressed by the mean effective pressure, and the temperature of the combustion air and therefore of the ambient air.

The spark should advance with an increase in speed of rotation and load, so that the combustion is complete the moment the expansion becomes effective; and the spark should retard with an increase in the temperature of the combustion air to prevent the temperature at the end of compression from reaching the limit at which cooling between two cycles

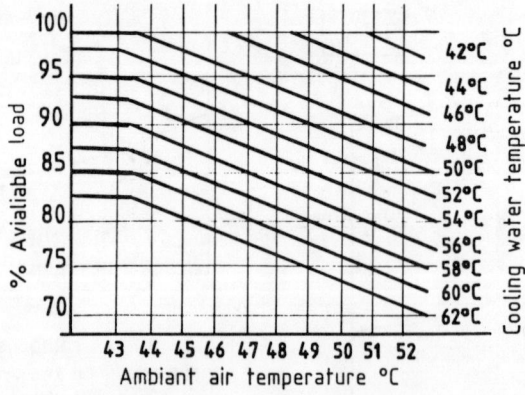

Figure 3-52. Influence of ambient air temperature on available power.

is insufficient and there is detonation. The spark advance is regulated by the timing of a circuit breaker in an ignition system. In ordinary operation the ignition advance should be regulated as a function of the speed and the load, a temperature correction not usually being made. The point of optimum ignition corresponds to the detonation limit.

When two-cycle engines are supercharged by a turbo-blower, however, the turbo-blower must be put into service so that the engine can be given air during startup. This is done either by mechanically driving the turbo-blower through a hydraulic coupling whose liquid is progressively removed as the exhaust gases accept the load of supplying power to the blower, or by injecting air to the turbine as long as the exhaust gases do not provide power for driving the blower. In this latter case, the engine is started with full ignition advance corresponding to the nominal torque; and right after the first few turns the ignition is retarded to the maximum ($25°-35°$ delay), so that the combustion occurs after top dead center and the exhaust gases are at a higher temperature, because of the reduced expansion, to supply maximum energy to the turbine driving the turboblower.

When fuel gas is injected to an engine through a valve with a fixed opening per cycle, the fuel gas pressure is a function of the speed and the mean effective pressure. The scavenging air pressure is then proportional to the pressure of the fuel gas to maintain a constant richness (Figure 3-53).

The ignition advance control during complete startup can be summarized as follows:

• At startup the ignition is at full advance.
• Right after the first few turns, the ignition is retarded to the minimum.

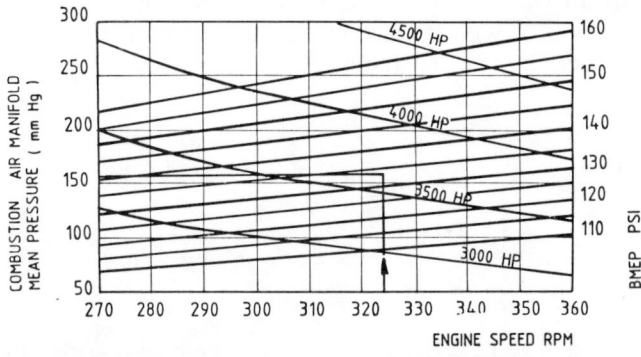

Figure 3-53. Combustion air pressure as a function of BMEP. (Courtesy of Cooper Industries.)

100 Drivers for Rotating Equipment

- As the engine picks up speed and power, the scavenging air pressure increases. This increased pressure acts on the pneumatic bellows of the ignition timing and advances it (Figure 3-54).
- The startup air (Figure 3-54) puts maximum pressure on the timing bellows to get a full advance.
- The startup air is cut off on detection of the threshold speed or threshold pressure of scavenging air, and the timing bellows exhausts across a regulating orifice (Figure 3-54) that slows the rate of change of ignition timing from full advance to full retard.
- The scavenging air pressure increases (the air charge to the turbocharger is cut off as soon as the scavenging air pressure reaches a threshold pressure), valve C (Figure 3-54) isolates the purge from the timing bellows, and the increase in scavenging air pressure advances the ignition.
- The ignition advances to its maximum with the scavenging air pressure at 75% of the pressure at nominal torque and speed, i.e., at 75% of mean effective pressure at nominal speed.

In other instances the ignition is regulated both by a signal representing the scavenging air pressure and a signal of speed, with an auto-selector between the two signals. At startup the pressure of the scavenging air acts alone, and the ignition advances after a full retard of 25°-30°. The signal for speed then takes over at 200-250 rpm, depending on the engine, and the ignition becomes a linear function of the rotating speed. On the Cooper Bessemer Quad, for example, ignition is 9° before top dead center at the nominal speed of 475 rpm and 4.5° at 300 rpm.

On engines operating according to the "ambient rating" system at over the nominal torque, a decrease in the temperature of the combustion air activates a signal enabling the torque to retard the ignition 2°-3° beyond nominal in order to avoid detonation.

By contrast to the gas engines described here, automotive gasoline engines are very fast, with mechanical ignition systems in which the spark is advanced both by (1) a centrifugal system in which little weights rotat-

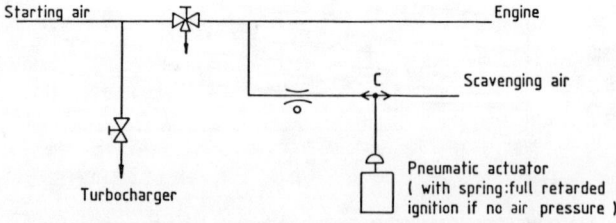

Figure 3-54. Action of the ignition control during startup.

ing inside the distributor advance the position of the rotor on its drive shaft as their centrifugal forces carry them away from the shaft, and (2) a timing bellows activated by pressure taken from after the carburetor's butterfly valve.

Control of the Air/Gas Ratio

The air/gas equivalence ratio (inverse of the richness) is determined by the manufacturer so as to avoid detonation at a volumetric ratio, nominal torque, and with the chosen fuel, as all these variables are brought together in the design. The fuel gas pressure is a function of the power demand (BMEP × rpm); the combustion air pressure should vary as a linear function of the fuel gas pressure to keep the air/gas ratio constant (Figure 3-55). The gas-to-air richness can be increased when the mean effective pressure (BMEP) decreases because the temperatures are lower and farther away from detonation (Figure 3-55).

In order to have the air/gas ratio automatically controlled, the fuel gas pressure is measured and the air pressure is kept proportional with a control system acting on a bypass of the supercharging turbine or on a butterfly valve in the air manifold.

Influence of the Fuel Quality

If the anti-detonating rating of the fuel gas drops, there is risk of detonation because the volumetric ratio cannot change; either the richness must be reduced or the ignition retarded so as to reduce the temperature.

If the heating value of the fuel increases, the richness and therefore the temperature increases. If the torque control system is based on the flow of fuel gas, this increase in heating value is translated to an increase in torque through an increase in the average effective pressure and the temperature. The increase in temperature leads to detonation; therefore the

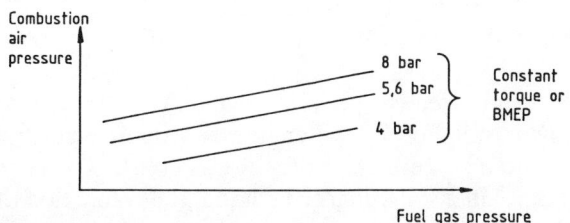

Figure 3-55. Variation of combustion air pressure with fuel gas pressure at constant BMEP.

richness must be reduced by increasing the combustion air pressure. It is thus important to have a good torque control system.

If the engine must operate with several different fuels having anti-detonating ratings close to each other but slightly different heating values, i.e., with natural gas from different sources, the air/gas ratio can be regulated for the richest fuel, which however brings on a lower thermal efficiency with the leanest fuel and can lead to faulty ignition. Another method is to reduce the ignition advance so as to avoid detonation. This latter method leads to a lower thermal efficiency no matter what the heating value, but it is the only possible method on automobile engines where reducing the richness leads to poor ignition.

It can be seen that the risk of detonation due to higher heating values of the fuel gas in gas engines increases as the average effective pressure is higher; lower rated engines with lower explosion pressures are much less sensitive to changes in the fuel gas, but of course they are bigger and cost more due to their greater cylinder capacity for equivalent power (although that does not mean their efficiency is lower).

Those engines operating with a precombustion chamber to reduce NO_x emissions generally operate with a lean mixture that delays the appearance of detonation at an equivalent volumetric ratio. This system also assures more even explosion pressures, a more regular thermal load per cylinder, and thus less risk of detonation.

Control of the Mean Effective Pressure (BMEP)

The nominal power of a gas engine, as defined by its nominal speed, is the maximum guaranteed by the manufacturer for continuous operation at a specified ambient temperature, where one allows for an increase in power at lower ambient temperatures. This nominal power corresponds to a mean effective pressure that should not be surpassed in order to protect the life of the machine, as well as to reduce maintenance costs and avoid detonation. For modern supercharged engines, the mean effective pressure limit is:

- 7.5–9 bar (110–130 psi) for two-cycle engines.
- 9.5–11.5 bar (140–170 psi) for four-cycle engines.

For a given mean effective pressure, which corresponds to a specific torque, the power supplied by the engine is proportional to the rotating speed. In fact, the mean effective pressure is constant only for a nonsupercharged engine; in a supercharged engine, it is a function of the supercharging, and thus of the load. An engine is supercharged when its mean effective pressure surpasses the value set by the manufacturer. Usually

Gas Engines   103

10% supercharge is allowed, but such operating conditions are not guaranteed by the manufacturer. They are associated with a more rapid wear and can be tolerated only in exceptional cases (for example, in the event that one unit in a compressor station breaks down) and only so long as detonation does not occur.

It should be noted that surpassing the nominal speed has less severe consequences than surpassing the torque and mean effective pressure. This is why torque control is so important.

When the power of a gas engine is to be reduced, as, for example, in reducing the flow of reciprocating compressors driven by gas engines, it is preferable to reduce the speed. A reduction in torque would be reflected in a lower thermal efficiency; the operation of the engine would be less regular, and there would be pollution of the lube oil by unburned fuel in the exhaust. The torque should never go below 60%.

Torque can be controlled by one of three methods:

1. Calculated from the operating characteristics of a driven reciprocating compressor.
2. Calculated torque based on the engine's cylinder pressure.
3. Calculated torque based on the flow of fuel gas.

Constant torque operating curves can be determined as functions of the suction and discharge pressures of reciprocating compressors (Figure 3-56), each curve corresponding to a given clearance-pocket volume relative to the cylinder volume (See Volume 1 of this series, *Compression Equipment*). Such plots are used with the measured suction and discharge pressures to identify that clearance pocket volume which will maintain the operation just below nominal torque. Regulation is done either with an instrument-pen making electrical contact with the field between two constant-torque curves on a printed circuit or with a computer that has the fields in its memory. This system is independent of any variation in fuel

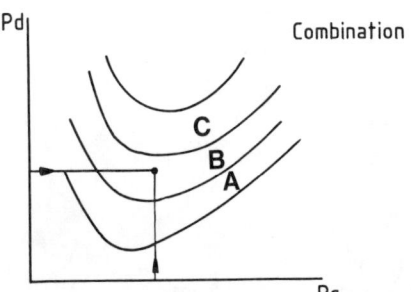

Figure 3-56. Constant-torque operating curves corresponding to specific clearance-pocket volumes.

gas quality but it can be used only if the driven machine is a reciprocating compressor with regulation by clearance volumes, and it becomes complicated as soon as the engine has a variable load due to changing air temperature.

In the system based on cylinder pressure, the pressure in each cylinder is measured by an electronic sensor connected to the cylinder head through a surge bottle and a valve (Figure 3-57). The surge bottle averages the pressure variation and gives the mean effective pressure. The measurements from all sensors are added electrically, the signal is corrected as a linear function of the rotating speed to account for the increase in mechanical friction with speed, and the resulting signal gives the value of the torque. This system is excellent and the best, but it applies only to two-cycle engines where all cycles are driving. It is independent of the fuel gas's heating value, as well as that of the driven machine. Nevertheless it should be complemented with a cyclic scanner of the individual engine cylinders' exhaust temperatures with comparison to the average of all the cylinders, so as to detect a poor ignition of any cylinder and stop the engine. If a cylinder does not fire, this system transfers that cylinder's load to the others through an increased mean effective pressure.

In the system based on the fuel gas flow rate the fuel gas pressure is maintained constant by a pressure-reducing valve. There is a surge tank between the gas flow measuring orifice and the gas pressure-control system at the inlet to the engine, as in all systems feeding fuel gas to engines. The gas flow measurement is by means of an orifice with pressure and temperature correction. Control is by one of three methods:

1. The pneumatic (or electronic) controller puts out a signal corresponding to $(p \, \Delta p/T)^{1/2}$ and representing the corrected flow in terms of pressure, p, and temperature, T, at the point of measurement, with Δp the differential pressure at the orifice.

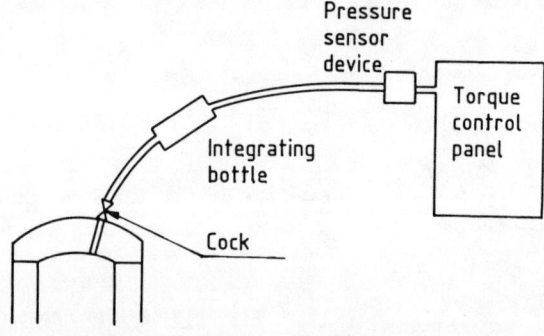

Figure 3-57. System for monitoring cylinder pressure.

2. The temperature, T, of the fuel gas is controlled to the most constant value while the pressure, p, is modified by a pressure controller in the measuring run as a function of temperature in order to maintain the ratio p/T constant while the calculator works out the signal $(\Delta p)^{1/2}$.
3. The temperature, T, and the pressure, p, of the fuel gas are controlled to the most constant value; the flow is measured in an orifice meter run and the calculator works out the signal $(\Delta p)^{1/2}$. This method is simpler.

This last system has the disadvantage of requiring a means for heating the startup fuel gas before expansion, whereas the first two permit startup while heating the fuel gas before expansion with cooling water circulating through the engine.

In all three cases the signal put out by the controller represents the normal flow of fuel gas and is divided by the rotating speed to give the mean effective pressure.

The system based on fuel gas flow rate has the disadvantages of not directly measuring the mean effective pressure and of not taking into account any variations in fuel gas quality. An increase in heating value is reflected by an increase in the mean effective pressure and risks detonation in heavy duty machines.

Maintenance

The maintenance of a gas engine includes different series of systematic and periodic visits, controls, and partial or complete overhauls. The details of all the different maintenance operations vary with the machine and are given in specialized manuals for each type of engine by each manufacturer. Some operations are routine, such as checking oil levels, lubricating, and cleaning filters, and can be done by either a mechanic or a mechanic's assistant. Others requiring higher qualifications are performed only by the mechanics; and these include the control and checking of the ignition, distribution, safety equipment, instrumentation, and torque-control systems. Finally, specialized teams directed by a foreman mechanic do complete reviews of the engines, dismantling the machine and checking for wear of such principal parts as cylinder heads, valves, linings, pistons and rings, bearings, bushings, and the turbocharger.

In sum, the maintenance of gas engines is continuous and requires competent personnel through which the gas engine's excellent ruggedness and operating safety are realized. Attention should be drawn to the importance of frequent control and regulation of the ignition, as well as of the air/gas ratio, and of operating without detonation or pre-ignition.

106 Drivers for Rotating Equipment

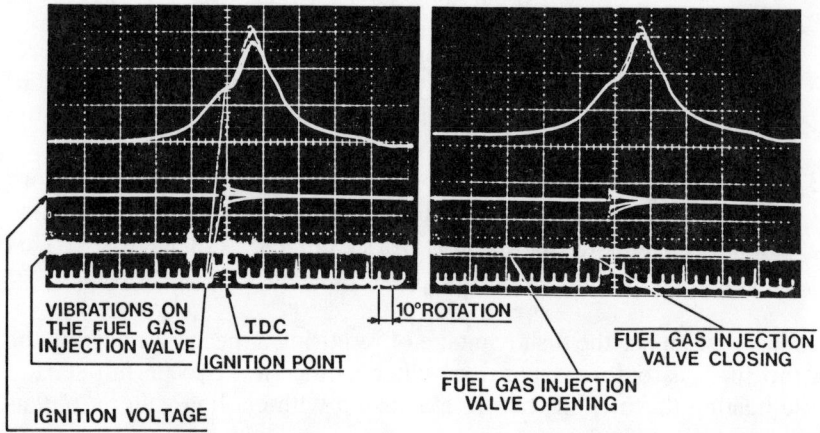

Figure 3-58. Two-cycle cylinder pressure time diagram.

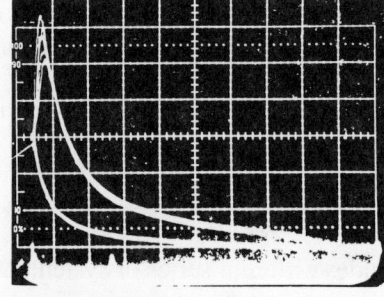

Figure 3-59. Two-cycle engine cylinder pressure diagram.

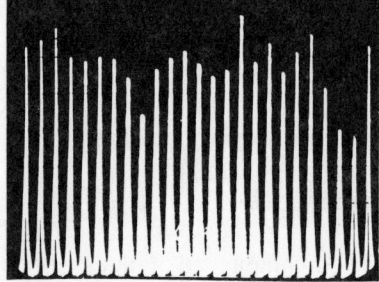

Figure 3-60. Maximum explosion pressure per cycle in a cylinder.

It is possible to carry out preventative maintenance through the systematic use of an electronic analyzer that:

1. Records the vibrations at different points in the engine and compares their amplitude and shapes to the readings made on a machine that was either new or recently rebuilt, in order to detect abnormal clearances and defective parts.
2. Records the characteristics of the ignition.
3. Records the indicator diagrams, which reveal both a loss of seal in valves or rings and poor ignition timing.

Examples of these recordings are shown in Figures 3-58, 3-59, and 3-60. Figure 3-58 shows the action of the gas injection valve and ignition with the cylinder pressure. Figure 3-59 shows a pressure volume diagram that reveals variations in pressure at the end of combustion, and an insufficient ignition advance. Figure 3-60 shows a plot of maximum cycle pressure for a series of cycles, which plot reveals that the maximum pressures are very irregular, due to poor regulation of the ignition and air/gas ratio.

Some typical maximum pressures after explosion are:

- 50 bar on a Clark TLA or TCV engine.
- 67 bar on a Cooper Bessemer GMVH engine with 100% torque.
- 74 bar on a Cooper Bessemer GMVH engine with 110% torque.

Chapter 4
Gas Turbines

Recent years have seen a considerable expansion in the industrial applications of gas turbines, particularly for gas transmission. When designed with two independent shafts for the air compressor and the output turbine, they offer a great flexibility of use. They combine a simple thermodynamic cycle with the enormous advantage of continuous rotating movement, but their evolution and success have been largely based on improvements in alloys with good mechanical and chemical resistances at high temperatures. They are currently used for applications requiring powers ranging from less than 1,000 kW to more than 75,000 kW. Taking account of current possibilities, their thermal efficiency based on LHV is somewhere between 16% and 36%, for simple installations without recovery of the exhaust heat, and between 26% and 34% for installations with exhaust heat recovery.

Thus the efficiency of gas turbines remains lower than that of gas engines, especially under partial load. Use of gas turbines with combined cycles does afford efficiencies of 40%–45%.

In gas transmission, the gas turbine shows many advantages: It is simple; its first cost for installation is low; it has wide flexibility; load changes and remote control are easy; and its needs for surveillance and maintenance are minimal. All these often make it the preferred driver in regions where water is scarce, in situations where maintenance and operating personnel are not on hand, or when the cost of fuel does not dominate the economics because the anticipated operating hours are limited or the fuel cost is low.

TECHNICAL DESCRIPTION OF A GAS TURBINE

A gas turbine can be characterized according to whether

- It has one or two independent shafts.
- It is a light turbine based on aerojet engine design, or a light industrial turbine derived from aerojet engines, or a heavy turbine derived from steam-turbine technology.

- Its thermal cycle is closed or open, simple or with recovery of exhaust heat, or combined.

and according to

- Its nominal performance (power, speed, thermal efficiency, combustion chamber outlet temperature, operation flexibility).

The characteristics of gas turbines are rated based on ambient conditions that are taken as:

- American NEMA conditions: 80°F and 1,000 feet of altitude (14.17 psia), equivalent to 26.7°C and 305 m altitude (0.977 bar absolute).

- ISO conditions: 15°C and 760 mm Hg (1.013 bar absolute). The ISO conditions are now generally used.

THERMODYNAMIC CYCLES

The basic thermodynamic cycle of a gas turbine was described by Brayton at the end of the 19th century. This Brayton cycle is analogous to that of steam turbines. However in the steamcycle (Rankine cycle) the fluid occurs as gas or liquid and undergoes vaporization and condensation during the cycle, whereas in the Brayton cycle the fluid is always gas. The two cycles go through four stages of operation as follows (Figure 4-1):

Step 1: The gas undergoes adiabatic compression in the Brayton cycle, whereas the pressure of liquid water is increased by a pump in the steam cycle.

Step 2: The temperature of the gas is increased at constant pressure through heat input in the Brayton cycle, whereas the water is vaporized and superheated by heat in the steam cycle.

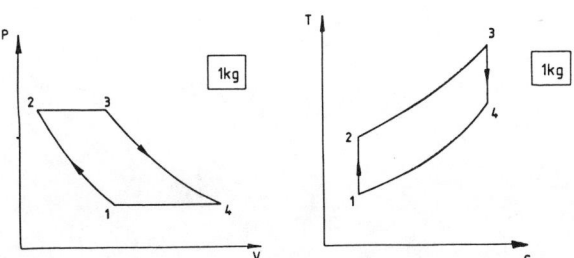

Figure 4-1. Theoretical gas turbine cycle diagrams.

110   Drivers for Rotating Equipment

Step 3: The gas undergoes adiabatic expansion in the Brayton cycle, and the superheated steam also undergoes adiabatic expansion in the steam cycle.

Step 4: The gas cools at constant pressure in the Brayton cycle, whereas the expanded steam is condensed at constant pressure in the steam cycle.

Within the limits of the thermal efficiency, the available power is proportional to the mass flow rate, which varies as: the compressor speed squared, a linear function of the compressor inlet pressure, the inverse of the compressor absolute inlet temperature. The difference between the power delivered by the expansion turbine (Step 3) and the power absorbed by the air compressor (Step 1) represents the available useful power. It corresponds roughly to one third the total power of the expansion turbine, and it increases with the efficiency. Typical of energy conversion cycles, the overall thermal efficiency is a function of:

- The compression ratio
- The inlet temperature to the expansion turbine
- The optimum compression ratio for any given inlet temperature to the expansion turbine
- The efficiency of the heat exchangers
- The combustion efficiency
- The adiabatic efficiency of the compressor and the expansion turbine.

Simple Closed-Cycle Gas Turbines

This system is shown in Figure 4-2. It involves four principal pieces of equipment, a compressor that is usually axial, a compressed gas heater, an expansion turbine, usually axial, and an exhaust gas cooler. This closed cycle system is rarely used, because the exchangers are expensive due to the low heat-transfer coefficients of the gases. However it does have the following important advantages:

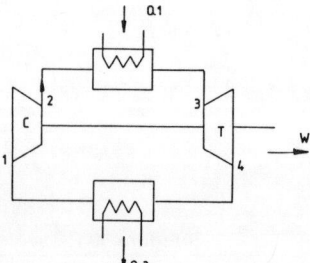

Figure 4-2. Simple closed-cycle gas turbine system.

- Any kind of fuel can be used, so that this system is possible with a nuclear reactor.
- There is less pollution.
- The gas is not polluted by the fuel and can be helium, argon, nitrogen, carbon dioxide, freon, or a petroleum gas to improve the heat transfer coefficients and reduce the temperatures.
- The system can be used with low temperature coolants (vaporizing liquified natural gas for example) that might freeze a liquid but not adversely affect a gas.
- The weight of the circulating gas and therefore the available power can be varied, which permits keeping the same speeds and maintaining the efficiency with constant temperatures.

Although the closed cycle is not widely used currently, it may see further development through use of recuperators or the addition of a steam cycle, as shown in Figures 4-3, 4-4, and 4-5.

Simple Open Cycle

In the open cycle, the primary heater of Figure 4-3 is replaced by a combustion chamber where liquid or gaseous fuel is injected directly, and

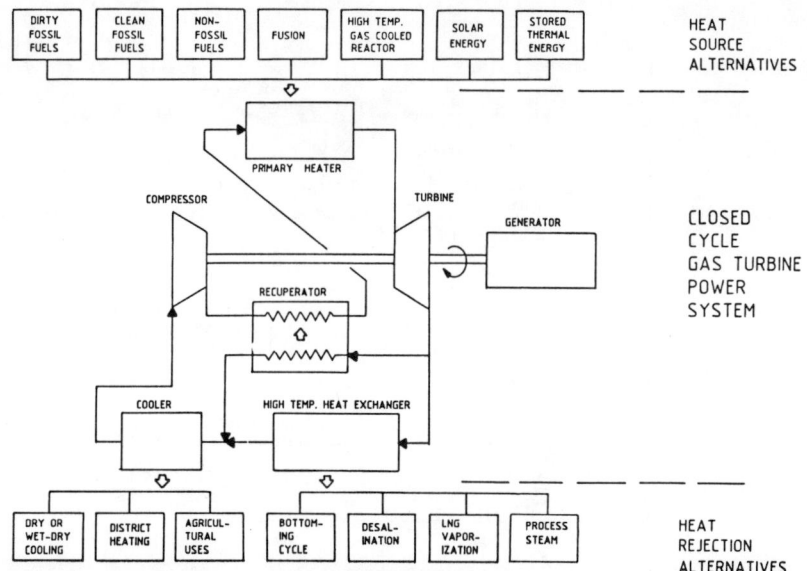

Figure 4-3. Alternate energy sources and exhaust heat uses for closed-cycle gas turbines. (Courtesy of the U.S. Department of Energy.)

112 Drivers for Rotating Equipment

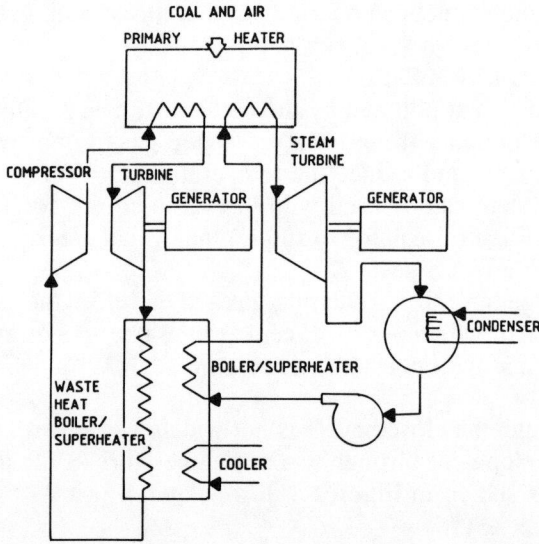

Figure 4-4. Steam-based closed cycle power system for gas turbines. (Courtesy of the U.S. Department of Energy.)

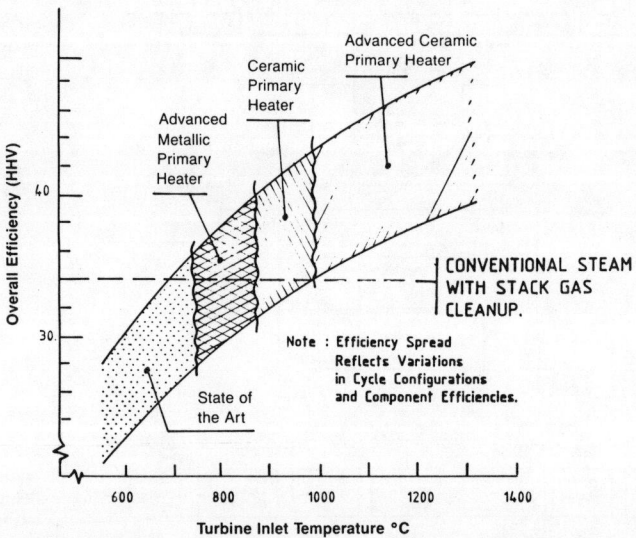

Figure 4-5. Projected phases in the development of closed-cycle gas turbines. (Courtesy of the U.S. Department of Energy.)

the heat exchanger is deleted, as the exhaust passes directly to the atmosphere from the outlet of the expansion turbine (Figure 4-6).

Actual Simple Cycle

Whether open or closed, the theoretical simple cycle must be modified in practice to take account for the adiabatic efficiencies of the compressor and turbine, for the pressure drop in the high-temperature exchanger or combustion chamber, and for the pressure drop in a possible low-temperature heat exchanger. These modifications change the cycle diagram of Figure 4-6 as shown in Figure 4-7. Consequently the theoretical thermo-

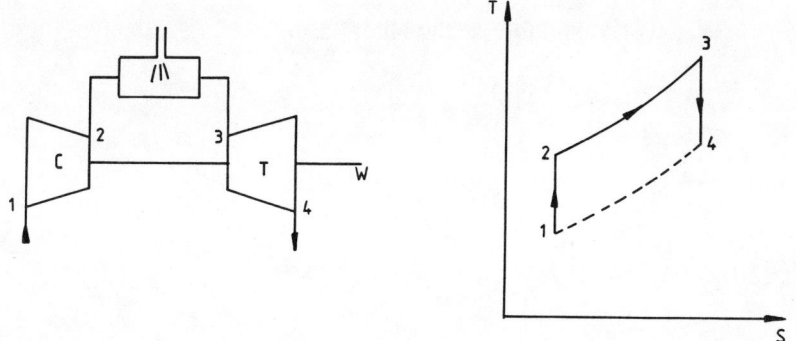

Figure 4-6. Theoretical flow scheme and cycle diagram for a simple cycle gas turbine.

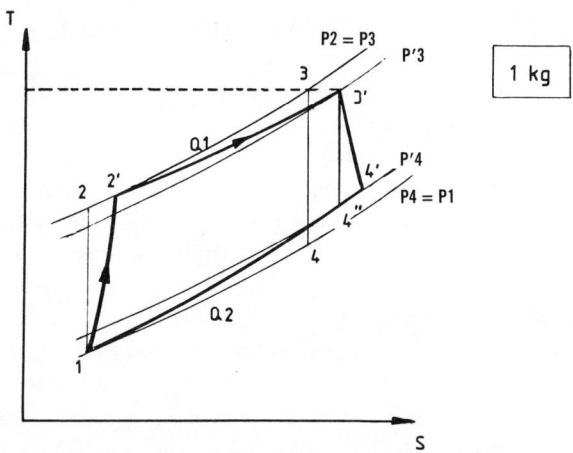

Figure 4-7. Modifications brought to the gas turbine cycle diagram by actual conditions and operations.

dynamic efficiency is reduced by a shape factor as defined in Steps 2′, 3′, and 4′ of Figure 4-7. The actual thermodynamic efficiency is

$$\eta_{actual} = \eta_{theoretical} \cdot \eta_{shape}$$

The theoretical thermodynamic efficiency can be calculated as:

$$\eta = \frac{W}{Q_1} = \frac{W_T - W_c}{Q_1} = \frac{(i_3 - i_4) - (i_2 - i_1)}{i_3 - i_2}$$

where W_c = work of the compressor
W_T = work of the expansion turbine
W = work available at the shaft

or

$$\eta = \frac{Q_1 - Q_2}{Q_1}$$

$$\eta = \frac{(i_3 - i_2) - (i_4 - i_1)}{i_3 - i_2}$$

$$\eta = 1 - \frac{i_4 - i_1}{i_3 - i_2}$$

where η = theoretical thermodynamic efficiency
i = enthalpy of the gas
1,2,3,4 = subscripts denoting conditions at the corresponding step in the cycle diagram (Figures 4-6 and 4-7).

Given the enthalpy diagram for the gas, one can determine i_1, i_2, i_3, and i_4 from the corresponding conditions at the inlets and outlets of the compressor and turbine. In the case of an open cycle the gas can be assumed as air, because the combustion is done with a very large excess of air due to temperature limitations of the materials. Thus $\gamma = c_p/c_v = 1.41$, and the theoretical efficiency for air when acting as a perfect gas is:

$$\eta = 1 - \frac{1}{\left(\frac{p_2}{p_1}\right)^{\frac{\gamma-1}{\gamma}}}$$

For an actual cycle,

η_c = the adiabatic efficiency of the compressor

η_T = the adiabatic efficiency for the expansion turbine

$\Delta p_2 = p_3' - p_3$

$p_3 = p_2$

$\Delta p_1 = p_4' - p_4$

$p_4 = p_1$

The work available from the shaft is $W = (W_T - W_c) = (i_3' - i_4') - (i_2' - i_1)$. The actual thermodynamic efficiencies can be written from the actual cycle diagram as follows:

$$\eta_{actual} = 1 - \frac{i_4' - i_1}{i_3' - i_2'} = 1 - \frac{T_4' - T_1}{T_3' - T_2'}$$

$$\eta_c = \frac{i_2 - i_1}{i_2' - i_1} = \frac{W_{c\,theoretical}}{W_{c\,actual}}$$

$$\eta_T = \frac{i_3' - i_4'}{i_3' - i_4''} = \frac{W_{T\,actual}}{W_{T\,theoretical}}$$

Then given the knowledge of the ambient pressure, p_1, the compressor's compression ratio, p_2/p_1, the inlet temperature to the turbine, $T_3' = T_3$, the adiabatic efficiencies, η_c and η_T, as well as the pressure losses Δp_1 and Δp_2, it is possible to determine i_1, i_2, i_3, and i_4 by means of an enthalpy diagram, and

$$i_2' = i_1 + \frac{i_2 - i_1}{\eta_c}$$

Also, i_3' corresponds to $p_2 - \Delta p_2$ and to T_3'; and i_4'' is determined from the isentropic expansion from $(p_2 - \Delta p_2)$ to $(p_1 + \Delta p_1)$. And:

$$i_4' = i_3' + \eta_T (i_4'' - i_3')$$

Drivers for Rotating Equipment

Since the air acts as a perfect gas:

$$\eta_c = \frac{\dfrac{\gamma}{\gamma-1} r\,(T_2 - T_1)}{\dfrac{\gamma}{\gamma-1} r\,(T_2' - T_1)} = \frac{\text{isentropic compression}}{\text{compression in a machine}}$$

$$\eta_T = \frac{\dfrac{\gamma}{\gamma-1} r\,(T_3' - T_4')}{\dfrac{\gamma}{\gamma-1} r\,(T_3' - T_4'')} = \frac{\text{expansion in a machine}}{\text{isentropic expansion}}$$

And since:

$$i_2' - i_1 = c_p\,(T_2' - T_1)$$

$$= \frac{\gamma}{\gamma-1} r\,(T_2' - T_1)$$

$$i_3' - i_4' = c_p\,(T_3' - T_4')$$

$$= \frac{\gamma}{\gamma-1} r\,(T_3' - T_4')$$

$$\eta_c = \frac{T_2 - T_1}{T_2' - T_1}$$

$$\eta_T = \frac{T_3' - T_4'}{T_3' - T_4''}$$

$$T_3' = T_3$$

$$\frac{T_2}{T_1} = \left(\frac{p_2}{p_1}\right)^{\frac{\gamma-1}{\gamma}} = r_c{}^\alpha$$

$$\alpha = \frac{\gamma-1}{\gamma}$$

$$\frac{T_3'}{T_4''} = \left(\frac{p_3'}{p_4'}\right)^{\frac{\gamma-1}{\gamma}} = \left(\frac{p_2 - \Delta p_2}{p_1 + \Delta p_1}\right)^{\frac{\gamma-1}{\gamma}} = r_T{}^\alpha$$

This leads to:

$$T_4' = T_3 - \eta_T T_3 \left(1 - \frac{1}{r_T^\alpha}\right)$$

$$T_2' = T_1 + \frac{T_1}{\eta_c}(r_c^\alpha - 1)$$

$$\eta = 1 - \frac{T_3 - T_1 - \eta_T T_3 \left(1 - \frac{1}{r_T^\alpha}\right)}{T_3 - T_1 - \frac{T_1}{\eta_c}(r_c^\alpha - 1)}$$

$$\eta = \frac{\eta_T \, \eta_c \frac{T_3}{T_1}\left(1 - \frac{1}{r_T^\alpha}\right) + 1 - r_c^\alpha}{\eta_c \left(\frac{T_3}{T_1} - 1\right) + 1 - r_c^\alpha}$$

$$\eta = \left(1 - \frac{1}{r_T^\alpha}\right)\left[\frac{\eta_T \, \eta_c \frac{T_3}{T_1} - r_T^\alpha \frac{r_c^\alpha - 1}{r_T^\alpha - 1}}{\eta_c \left(\frac{T_3}{T_1} - 1\right) + 1 - r_c^\alpha}\right]$$

Example

Calculations for a turbine with a simple open cycle at ISO conditions. Because the gas is air

$$\gamma = 1.41$$

$$\alpha = \frac{\gamma - 1}{\gamma} = 0.291$$

Assume that:

$T_3 = 1{,}273 \text{ K} = 1{,}000°C$

$T_1 = 288 \text{ K} = 15°C$

and the altitude is sea level.

118   Drivers for Rotating Equipment

We have seen that maximum power can be obtained at a compression ratio, r, as follows:

$$r = \left(\frac{T_3}{T_1}\right)^{\frac{\gamma}{2(\gamma-1)}}$$

Accordingly the maximum power is assumed for a compression ratio, r_c, of:

$$r_c = \left(\frac{T_3}{T_1}\right)^{\frac{\gamma}{2(\gamma-1)}} = 4.42^{1.72} \simeq 12.88 \simeq 13$$

The thermodynamic efficiency for a theoretical cycle is:

$$\eta = 1 - \frac{1}{(r_c)^{\frac{\gamma-1}{\gamma}}} = 0.526$$

For the actual cycle, we can assume that the adiabatic efficiencies of the compressor, $\eta_c = 0.84$, and the turbine, $\eta_T = 0.88$; and $(\Delta p_2/p_2) = 4\%$. Thus . . .

$$r_T = \frac{1.013 \times 13 \times 0.96}{1.013} \simeq 12.48$$

The thermodynamic efficiency is therefore . . .

$$\eta = \left(1 - \frac{1}{12.48^{0.291}}\right)\left[\frac{0.84 \times 0.88 \times 4.42 - 12.48^{0.291}\dfrac{13^{0.291}-1}{12.48^{0.291}-1}}{0.84(4.42-1)+1-13^{0.291}}\right]$$

$$\eta = 0.335$$

And the shape efficiency, $\eta_s = 0.335/0.526 = 0.637$.

The overall thermal efficiency, R_g, takes into account not only the thermodynamic efficiency determined from the actual cycle, but also the losses due to incomplete combustion, η_{co}, heat losses from the combustion chamber, the heat exchanger (in a closed cycle) and the mechanical losses, η_m. Thus . . .

$$R_g = \eta_{act} \times \eta_{co} \times \eta_m$$

Gas Turbines

The combustion and mechanical efficiencies improve as the power of the machine approaches the maximum for a given flow rate of gas, i.e., when the compression ratio, r_c, corresponds to $(T_3/T_1)^{1/2\alpha}$. For each value of (T_3/T_1) there is therefore a compression ratio leading to a maximum overall efficiency. Figure 4-8 gives a specific example of the variation of the overall thermal efficiency when the inlet temperatures to the expansion turbine are 700°C and 770°C.

Maximum Allowable Temperatures at the Inlet to the Turbine:

The maximum temperature for a gas turbine depends on principally the metals or alloys that have been selected, the length of life fixed for the machine, the service whether continuous or intermittent at full load, the thermodynamic design of the first impeller, and the cooling of exposed parts.

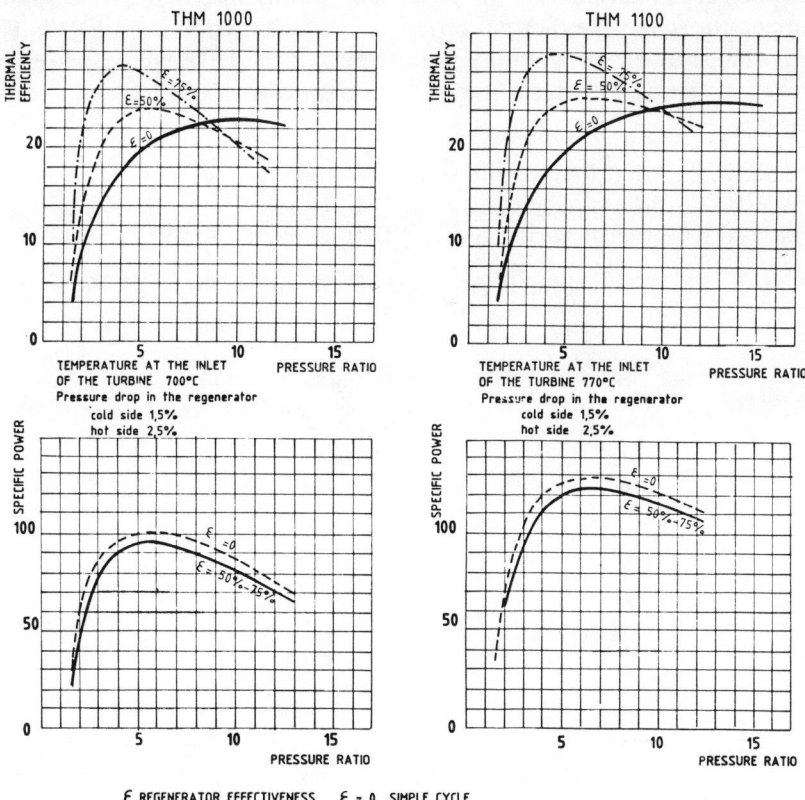

Figure 4-8. Influence of regenerator efficiency and turbine pressure ratios on thermal efficiency and specific power of gas turbines (Courtesy of Hispano Suiza.)

120 Drivers for Rotating Equipment

Those metals and alloys chosen for gas turbines are chosen primarily on the basis of their performance at high temperatures, as expressed in terms of their elastic limits and their creep strengths at temperatures typical of those in the turbine. The design of the first impeller and the cooling of exposed parts are related through designs that draw off cool air from the compressor and use that to cool critical parts of the first stages of the turbine, as well as designs that use holes to improve cooling of the impeller fins of the first stages and create an air film on the surface. Thus it is currently possible to reach temperatures of 1170°C at the outlet of the combustion chamber in industrial turbines, while holding skin temperatures to 820°C, whereas in 1968, the maximum temperatures at the outlet of the combustion chamber were 700°–800°C in industrial machines with peaks of 900°–950°C in aircraft gas turbines at the time of take-off. Also, the thermal efficiencies were 16%–25% in 1968, whereas current turbines reach efficiencies of 36%. The trends in compression ratios, temperatures and the thermal efficiencies are shown in Figure 4-9. The importance of the temperature at the inlets of the first expansion impellers is shown by its effect on the life of a Nimonic 105, which is as follows:

- At 850°C, the projected life is 100,000 hours.
- At 950°C, the projected life is 10,000 hours.
- At 1200°C, the projected life is 3,000 hours.

Each startup of a turbine corresponds to about 10 hours of life.

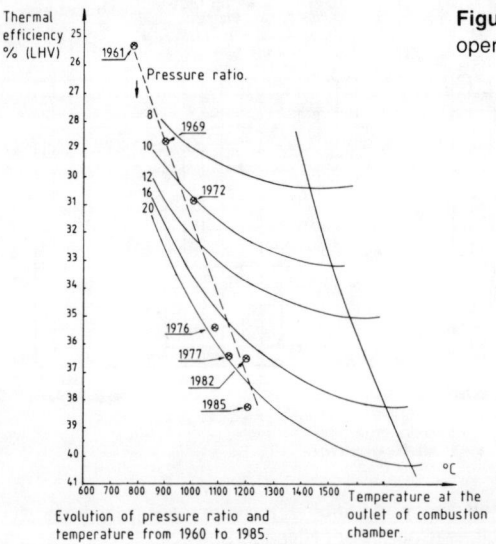

Figure 4-9. Progressive development of operating characteristics of gas turbines.

Open Cycles with Heat Recovery

The temperature of the exhaust gas, which is a function of the combustion temperature and the compression ratio, is already rather high to permit a heat recovery. Because the flow of exhaust gases is nearly equal to the flow of compressed gases (considering the low flow of fuel), and because their specific heats are practically the same, it is theoretically possible to raise the temperature of the compressed gas to that of the gases at the outlet of the turbine, by means of heat exchange. Such a system is shown schematically in Figure 4-10 for an open cycle in which an entropy diagram is drawn for one kg of air flow without taking into account the pressure drop in the combustion chamber. A turbine using such a cycle costs 20%–30% more than a simple-cycle turbine.

Examining the cycle one can see that recovering the heat of the exhaust gases is possible only if T_4 is higher than T_2. The difference between T_4 and T_2 decreases for a given combustion temperature, T_3, as the compression ratio increases; and that difference can be cancelled out, or T_4 can even become less than T_2 at high compression ratios. Considering this cycle therefore suggests two alternatives: Either use a high compression ratio to obtain a high efficiency from the simple cycle, with heat recovery not being of interest because of its cost and limited improvement in efficiency, or have a moderate compression ratio with poor efficiency of the simple cycle and use an exchanger to achieve an efficiency equivalent to, if not better than, the first case.

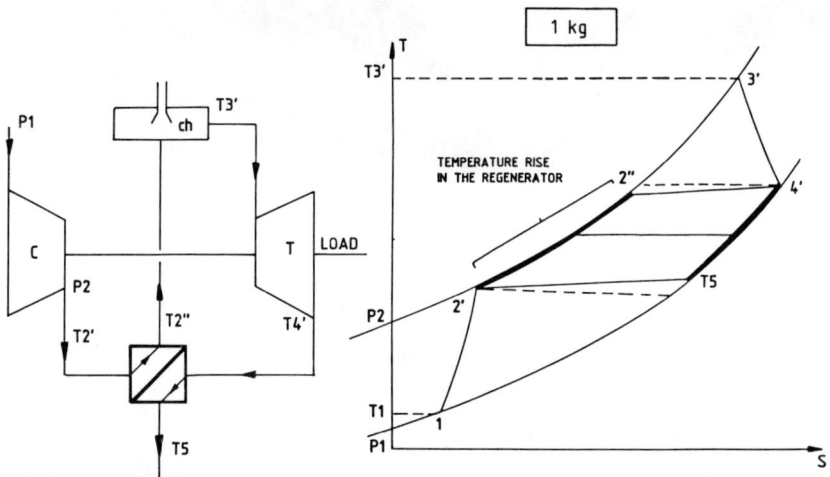

Figure 4-10. Flow scheme and cycle diagram for the open cycle gas turbine with heat recovery.

122 Drivers for Rotating Equipment

Whether or not there is a recuperator, it is always important to have the combustion temperature as high as possible.

The disadvantage of the heat recovery unit is that it introduces a large thermal mass that prevents the system from handling rapid variations in temperature. Therefore the heat recovery unit is not recommended when a turbine is to be stopped and started frequently or when the load is to be changed often. The heat exchanger makes it obligatory to stop and start the turbine through ascending or descending phases of low power, whereas simple-cycle turbines can accept full load within a few minutes and in some cases a few seconds, as for example in electrical-generating back-up groups.

Some heavy heat exchangers with the exhaust gas passing through tube bundles and the compressed air passing across the tubes between shell baffles also give good results with rapid variations in temperature. Newer heat exchangers with sheet-metal transfer surfaces are designed to have low thermal mass in order to accept temperature variations equivalent to those possible in flexible turbines during normal startup. The design of these exchangers is such that the exhaust gases and compressed air are kept separated by sheet metal that has no braised edges and is free to move in relation to the fixed structure (Figure 4-11). These heat ex-

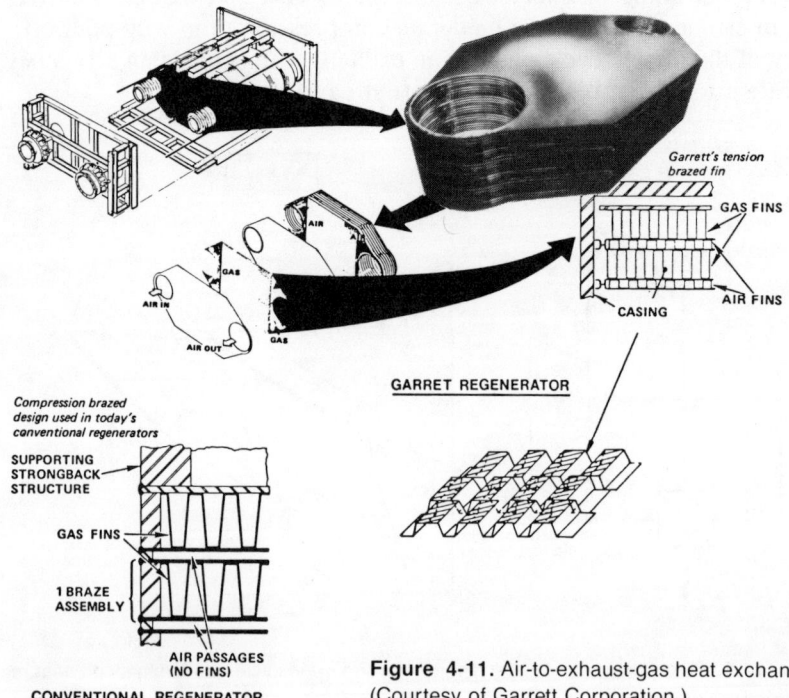

Figure 4-11. Air-to-exhaust-gas heat exchangers. (Courtesy of Garrett Corporation.)

changers can recover 83%-88.9% of the available heat with a reduced pressure drop, as follows:

Efficiency:	85.1%	88.5%
Pressure drop:		
Of the exhaust gas:	0.8%	1.3%
Of the air:	0.7%	0.7%
Total:	1.5%	2.0%

Thermal Efficiency of Turbines with Heat Recovery

In addition to the compression ratio and the ratio of the extreme temperatures T_3/T_1, the thermal efficiency of heat recovery turbines is affected by the efficiency of the exchanger, ϵ, which is expressed as (Tables 4-1 and 4-2):

$$\epsilon = \frac{c_p (T_4' - T_5)}{c_p (T_4' - T_2')} = \frac{T_2'' - T_2'}{T_4' - T_2'}$$

These temperatures are identified in Figure 4-10, T_5 being higher than T_2' and T_2'', and less than T_4'. The efficiency of these exchangers is 0.80–0.90 of the available heat, at drops of 1.7–2.0% on the exhaust and air sides, respectively. The maximum inlet temperature is usually 550°–600°C.

Using the same terms as for simple-cycle gas turbines, the efficiency of turbines with heat recovery is determined as follows:

$$\eta = \frac{W}{Q_1} = \frac{W_T - W_c}{Q_1}$$

$$\eta = \frac{i_3' - i_4' - (i_2' - i_1)}{i_3' - i_2''}$$

$$\eta = \frac{T_3 - T_2' - (T_4' - T_1)}{T_3 - T_2' - (T_2'' - T_2')}$$

$$\eta = \frac{T_3 - T_2' - (T_4' - T_1)}{T_3 - T_2' - \epsilon (T_4' - T_2')}$$

(text continued on page 129)

Table 4-1
Enthalpy of Air, kcal/kg Above 0 K

P(bar)	T(K) 85	90	100	110	120	140	160
1	19.8	21.0	23.5	25.96	28.39	33.24	38.07
5	−26.98	−24.69	21.7	24.47	27.15	32.34	37.38
10	−26.91	−24.64	−19.9	22.2	25.38	31.12	36.45
20	−26.8	−24.50	−19.8	−14.8	20.4	28.32	34.49
30	−26.62	−24.4	−19.7	−14.8	− 9.27*	24.9	32.34
40	−26.48	−24.23	−19.6	−14.8	− 9.51	20.0	29.95
50	−26.31	−24.09	−19.6	−14.8	− 9.68	11.2	27.34
60	−26.17	−23.97	−19.5	−14.8	− 9.75	5.45	24.3
70	−26.00	−23.8	−19.3	−14.7	− 9.85	3.27	21.5
80	−25.86	−23.7	−19.2	−14.7	− 9.87	2.15	18.7
90	−25.69	−23.5	−19.1	−14.6	− 9.89	1.41	16.5
100	−25.55	−23.4	−19.0	−14.5	− 9.87	0.932	14.7
125	−25.14	−23.0	−18.7	−14.3	− 9.78	0.359	12.2
150	−24.71	−22.6	−18.3	−14.1	− 9.58	0.120	10.8
175	−24.31	−22.2	−18.0	−13.7	− 9.37	0.048	10.1
200	−23.9	−21.8	−17.6	−13.4	− 9.08	0.048	9.61
300	−22.1	−20.1	−16.0	−11.9	− 7.74	0.741	9.37
400	−20.3	−18.3	−14.3	−10.3	− 6.19	1.98	10.1
500		−16.5	−12.5	− 8.48	− 4.47	3.44	11.3

P(bar)	180	200	220	240	260	270	280
1	42.90	47.73	52.53	57.34	62.1	64.55	66.97
5	42.33	47.25	52.13	57.00	61.85	64.3	66.70
10	41.61	46.65	51.65	56.57	61.49	63.96	66.39
20	40.08	45.43	50.64	55.73	60.78	63.29	65.77
30	38.53	44.2	49.64	54.90	60.08	62.64	65.18
40	36.88	42.95	48.64	54.09	59.39	62.00	64.58
50	35.18	41.71	47.63	53.25	58.70	61.35	63.98
60	33.3	40.45	46.65	52.44	58.03	60.72	63.42
70	31.69	39.17	45.67	51.65	57.4	60.13	62.88
80	29.95	37.93	44.72	50.88	56.71	59.53	62.33
90	28.27	36.73	43.78	50.14	56.09	58.99	61.81
100	26.72	35.56	42.88	49.40	55.50	58.41	61.28
125	23.5	32.96	40.80	47.70	54.09	57.1	60.11
150	21.4	30.88	39.00	46.22	52.8	55.97	59.0
175	20.0	29.28	37.55	44.9	51.72	54.95	58.1
200	19.1	28.11	36.38	43.88	50.79	54.06	57.26
300	17.8	26.1	33.9	41.39	48.45	51.84	55.16

(table continued on next page)

Table 4-1
Continued

P(bar)	T(K)						
	180	200	220	240	260	270	280
400	18.1	25.93	33.48	40.75	47.73	51.12	54.44
500	19.1	26.62	33.9	41.04	47.92	51.29	54.59
600		27.7	34.92	41.90	48.66	51.98	55.26
700			36.16	43.04	49.74	53.03	56.28
800			37.62	44.43	51.07	54.32	57.55
900				45.94	52.53	55.78	58.99
1000					54.11	57.34	60.51

	290	300	320	350	400	500	600
1	69.36	71.77	76.58	83.82	95.89	120.3	145.2
5	69.12	71.56	76.38	83.7	95.79	120.3	145.2
10	68.8	71.29	76.17	83.46	95.65	120.2	145.1
20	68.26	70.7	75.72	83.10	95.38	120.1	145.1
30	67.71	70.22	75.24	82.72	95.1	120.0	145.1
40	67.2	69.72	74.8	82.36	94.9	119.9	145.0
50	66.61	69.21	74.38	82.02	94.62	119.9	145.0
60	66.01	68.70	73.93	81.69	94.42	119.6	145.0
70	65.58	68.26	73.54	81.36	94.2	119.5	145.0
80	65.32	67.80	73.1	81.0	93.95	119.5	145.0
90	64.58	67.33	72.75	80.71	93.74	119.4	145.0
100	64.12	66.90	72.37	80.42	93.52	119	145.0
125	63.02	65.89	71.51	79.71	93.04	119.1	145.0
150	62.02	64.96	70.72	79.09	92.61	118.9	145.0
175	61.14	64.15	70.00	78.51	92.23	119	145.0
200	60.40	63.45	69.41	78.03	91.90	118.7	145
300	58.41	61.59	67.76	76.70	91.04	118.5	145.4
400	57.69	60.90	67.2	76.26	90.8	118.7	146
500	57.8	61.04	67.30	76.46	91.13	119.3	146.8
600	58.48	61.69	67.95	77.10	91.82	120.1	147.8
700	59.49	62.67	68.90	78.06	92.80	121.2	149.0
800	60.73	63.88	70.12	79.25	94.00	122.4	150.3
900	62.1	65.29	71.48	80.59	95.4	123.8	151.7
1000	63.67	66.80	72.99	82.07	96.82	125.3	153.2

	700	800	900	1000	1100	1200	1300
1	170.6	196.6	223.1	250.14	277.65	305.54	333.79
5	170.6	196.6	223.2	250.21	277.69	305.61	333.86
10	170.6	196.6	223.2	250.28	277.77	305.7	333.93

(table continued on next page)

Table 4-1
Continued

P(bar)	T(K)						
	700	800	900	1000	1100	1200	1300
20	171	197	223.3	250.40	277.93	305.85	334.1
30	170.7	196.8	223.4	250.52	278.08	306.02	334.31
40	170.7	196.9	223.5	250.66	278.24	306.21	334.50
50	170.7	197	223.6	250.81	278.39	306.37	334.67
60	170.8	197.0	223.8	251.0	278.55	306.54	334.86
70	170.8	197.1	223.9	251.07	278.72	306.73	335.05
80	170.9	197.2	224.0	251.21	278.89	306.90	335.27
90	170.9	197.3	224.1	251.36	279.03	307.09	335.46
100	171.0	197.4	224.2	251.52	279.20	307.26	335.65
125	171.1	197.6	224.5	251.88	279.6	307.74	336.15
150	171.2	197.8	224.9	252.26	280.06	308.19	336.66
175	171.4	198.1	225.2	252.65	280.49	308.67	337.16
200	171.6	198.3	225.5	253.03	280.92	309.15	337.66
300	172.3	199.4	226.9	254.61	282.55	311.06	339.71
400	173.2	200.6	228.3	256.2	284.48	312.99	341.8
500	174.3	201.9	229.7	257.83	286.25	314.91	343.80
600	175.4	203.2	231.2	259.43	288.0	316.79	345.81
700	176.7	204.5	232.6	261.08	289.74	318.66	347.79
800	178	206	234	262.73	291.48	320.5	349.73
900	179.5	207.5	235.8	264.41	293.3	322.34	351.66
1000	181.1	209.1	237.5	266.08	295.02	324.18	353.58

- Passage from the liquid state to the vapor state.
* Critical point.
Courtesy of L'air Liquide and Elsevier Science Publishers B.V.

Table 4-2
Entropy of Air Above 0 K at 1,013 bar abs., kcal/kg
(Can be used for the gas turbine exhaust gas)

P(bar)	T(K)						
	85	90	100	110	120	140	160
1	1.336	1.35	1.376	1.400	1.421	1.459	1.491
5	0.7512	0.777	1.254	1.281	1.304	1.344	1.377
10	0.7502	0.7758	0.8250	1.218	1.246	1.290	1.326
20	0.7500	0.7741	0.8226	0.8697	1.167 *	1.23	1.270

(table continued on next page)

Table 4-2
Continued

	T(K)						
P(bar)	85	90	100	110	120	140	160
30	0.7471	0.7724	0.8205	0.8664	0.9151	1.182	1.232
40	0.746	0.7708	0.8183	0.8635	0.9099	1.134	1.201
50	0.7442	0.7691	0.8162	0.8606	0.9056	1.064	1.173
60	0.7428	0.7674	0.8143	0.858	0.9017	1.02	1.147
70	0.7414	0.7660	0.8124	0.856	0.8982	1.000	1.121
80	0.7399	0.7643	0.8104	0.853	0.8951	0.9890	1.099
90	0.739	0.7629	0.8085	0.8511	0.8934	0.9809	1.080
100	0.7373	0.7615	0.8069	0.8489	0.8893	0.9742	1.065
125	0.7340	0.7579	0.8026	0.8439	0.8833	0.9622	1.041
150	0.7309	0.7545	0.7987	0.8394	0.8778	0.9531	1.025
175	0.7278	0.7512	0.7952	0.8351	0.8728	0.9457	1.013
200	0.7261	0.748	0.7916	0.8310	0.8685	0.9395	1.003
300	0.7141	0.7368	0.779	0.8176	0.8537	0.9199	0.978
400	0.7048	0.7270	0.7686	0.8064	0.8420	0.9053	0.9598
500		0.7182	0.7593	0.7968	0.8322	0.8946	0.9479
	180	200	220	240	260	270	280
1	1.519	1.545	1.568	1.589	1.608	1.617	1.626
5	1.407	1.433	1.456	1.48	1.497	1.51	1.515
10	1.356	1.383	1.407	1.428	1.448	1.457	1.466
20	1.30	1.331	1.356	1.378	1.398	1.407	1.417
30	1.268	1.298	1.324	1.347	1.368	1.378	1.387
40	1.242	1.27	1.301	1.325	1.346	1.356	1.365
50	1.220	1.254	1.282	1.307	1.328	1.34	1.35
60	1.20	1.237	1.266	1.292	1.314	1.32	1.334
70	1.182	1.221	1.252	1.278	1.301	1.311	1.321
80	1.165	1.207	1.239	1.266	1.290	1.300	1.311
90	1.1499	1.195	1.228	1.256	1.280	1.29	1.301
100	1.136	1.183	1.218	1.246	1.270	1.282	1.292
125	1.108	1.157	1.195	1.225	1.250	1.26	1.273
150	1.087	1.137	1.176	1.207	1.233	1.245	1.256
175	1.071	1.120	1.160	1.192	1.219	1.231	1.243
200	1.059	1.107	1.146	1.179	1.206	1.22	1.231
300	1.027	1.07	1.108	1.141	1.169	1.182	1.194
400	1.007	1.048	1.084	1.116	1.144	1.157	1.17
500	0.9921	1.032	1.067	1.098	1.125	1.138	1.150
600		1.019	1.053	1.083	1.110	1.123	1.135
700			1.042	1.072	1.098	1.111	1.123

(table continued on next page)

Table 4-2
Continued

P(bar)	T(K)						
	180	200	220	240	260	270	280
800			1.032	1.062	1.088	1.100	1.112
900				1.053	1.079	1.091	1.103
1000					1.071	1.083	1.095
	290	300	320	350	400	500	600
1	1.634	1.642	1.658	1.679	1.711	1.766	1.811
5	1.523	1.531	1.547	1.569	1.601	1.656	1.701
10	1.474	1.483	1.50	1.520	1.553	1.608	1.653
20	1.425	1.434	1.450	1.472	1.505	1.560	1.605
30	1.40	1.404	1.420	1.443	1.476	1.531	1.577
40	1.37	1.383	1.399	1.42	1.46	1.511	1.557
50	1.357	1.366	1.383	1.406	1.439	1.495	1.541
60	1.34	1.352	1.369	1.392	1.426	1.482	1.529
70	1.331	1.340	1.357	1.380	1.415	1.471	1.52
80	1.320	1.329	1.347	1.370	1.405	1.462	1.508
90	1.311	1.320	1.337	1.361	1.396	1.45	1.500
100	1.302	1.311	1.33	1.353	1.388	1.445	1.492
125	1.283	1.293	1.311	1.335	1.371	1.429	1.476
150	1.267	1.277	1.296	1.320	1.357	1.415	1.463
175	1.253	1.264	1.282	1.308	1.345	1.404	1.452
200	1.242	1.252	1.271	1.297	1.334	1.394	1.442
300	1.205	1.216	1.236	1.263	1.301	1.36	1.412
400	1.180	1.191	1.211	1.238	1.277	1.340	1.389
500	1.161	1.172	1.192	1.220	1.259	1.322	1.372
600	1.146	1.157	1.177	1.205	1.244	1.31	1.358
700	1.134	1.14	1.165	1.192	1.232	1.295	1.35
800	1.12	1.134	1.154	1.181	1.221	1.284	1.335
900	1.14	1.125	1.14	1.172	1.211	1.275	1.326
1000	1.106	1.117	1.136	1.164	1.203	1.27	1.318
	700	800	900	1000	1100	1200	1300
1	1.851	1.885	1.917	1.945	1.971	1.995	2.018
5	1.74	1.775	1.806	1.835	1.861	1.885	1.908
10	1.692	1.727	1.758	1.787	1.801	1.837	1.860
20	1.645	1.679	1.711	1.739	1.765	1.790	1.812
30	1.617	1.65	1.683	1.71	1.74	1.762	1.785
40	1.60	1.631	1.663	1.691	1.718	1.742	1.765
50	1.581	1.616	1.647	1.676	1.702	1.727	1.749
60	1.568	1.603	1.63	1.66	1.69	1.714	1.737

(table continued on next page)

Table 4-2
Continued

P(bar)	T(K)						
	700	800	900	1000	1100	1200	1300
70	1.558	1.593	1.624	1.653	1.679	1.703	1.726
80	1.548	1.583	1.615	1.644	1.670	1.694	1.717
90	1.540	1.58	1.607	1.635	1.662	1.686	1.71
100	1.532	1.568	1.599	1.628	1.654	1.679	1.70
125	1.516	1.552	1.584	1.613	1.639	1.663	1.686
150	1.504	1.539	1.571	1.600	1.626	1.651	1.673
175	1.492	1.528	1.560	1.589	1.615	1.640	1.663
200	1.483	1.518	1.550	1.579	1.61	1.630	1.653
300	1.453	1.49	1.521	1.551	1.58	1.602	1.625
400	1.431	1.468	1.500	1.530	1.557	1.582	1.605
500	1.414	1.451	1.484	1.514	1.541	1.57	1.589
600	1.400	1.437	1.470	1.500	1.527	1.552	1.576
700	1.388	1.426	1.459	1.488	1.516	1.541	1.564
800	1.378	1.415	1.449	1.478	1.506	1.531	1.555
900	1.369	1.406	1.441	1.470	1.498	1.52	1.546
1000	1.361	1.40	1.431	1.461	1.489	1.515	1.538

- Passage from the liquid state to the vapor state.
* Critical point.
Courtesy of L'Air Liquide and Elsevier Science Publishers B.V.

(text continued from page 123)

$$\eta = \frac{\left(1 - \frac{1}{r_T^\alpha}\right)}{1 - \epsilon} \left[\frac{\eta_c\, \eta_T\, \frac{T_3}{T_1} - r_T^\alpha\, \frac{r_c^\alpha - 1}{r_T^\alpha - 1}}{\eta_c\left(\frac{T_3}{T_1} - 1\right) + 1 - r_c^\alpha + \frac{\epsilon}{1-\epsilon}\frac{T_3}{T_1}\eta_T\,\eta_c\left(1 - \frac{1}{r_T^\alpha}\right)} \right]$$

Example 1

$T_3 = 1273$ K

$T_1 = 288$ K

$p_1 = 1.013$ bar

$\gamma = 1.41$

$\alpha = 0.291$

130 Drivers for Rotating Equipment

$$\frac{T_3}{T_1} = 4.42$$

$$r_c = 13$$

$$\eta_c = 0.84$$

$$\eta_T = 0.88$$

$$\left.\begin{array}{l}\dfrac{\Delta p_2}{P_2} = 4\% + 2\% \\[2ex] \dfrac{\Delta p_1}{P_1} = 2\%\end{array}\right\} \text{Pressure drop of the recuperator: 2\% at the exhaust and 2\% at the inlet}$$

$$\epsilon = 0.8$$

$$r_T = \frac{1.013 \times 13 \times 0.94}{1.013 \times 1.02} = 11.98$$

$$T_4' = 696 \text{ K} = 423\,°\text{C}$$

$$T_2' = 668 \text{ K} = 395\,°\text{C}$$

Because $\eta = 0.337$ with heat exchange, compared to 0.335 as calculated earlier, there is no advantage in adding a heat exchanger.

Example 2

Instead of taking a compression ratio of 13, assume a machine with a compression ratio of 6:

$$r_T = \frac{1.013 \times 6 \times 0.94}{1.013 \times 1.02} = 5.53$$

$$T_4' = 833 \text{ K} = 560\,°\text{C}$$

$$T_2' = 522 \text{ K} = 249\,°\text{C}$$

Thus the efficiency without heat recovery is $\eta = 0.27$, while the efficiency with heat recovery is $\eta = 0.41$. Comparing the power of these machines for an air flow of 1 kg/sec:

- First machine without heat recovery:

    $T_3 = 1273$ K

    $T_2' = 668$ K

    $i_3 = 272.5$ kcal/kg

    $i_2' = 109.05$

    $i_3 - i_2' = 163.45$

The power is $0.335 \times 4.184 \times 163.45 = 229.1$ kW
- Second machine without heat recovery:

    $T_3 = 1273$ K

    $T_2' = 522$ K

    $i_3 = 272.5$

    $i_2' = 72.42$

    $i_3 - i_2' = 200.08$

The power is $0.27 \times 4.184 \times 200.08 = 226$ kW

- Second machine with heat recovery:

    $T_3 = 1273$ K

    $T_2'' = 522$ K $+ 0.8\ (833 - 522)$

    $T_2'' = 770$ K

    $i_3 = 272.5$

    $i_2'' = 135.4$

    $i_3 - i_2'' = 137.1$

The power is $0.41 \times 4.184 \times 137.1 = 235$ kW

The second machine without heat recovery has a lower power than the first machine because the compression ratio is not optimum for the combustion chamber outlet temperature. Because of the high efficiency of the heat exchanger, the second machine with heat recovery is more powerful than the first. The curves of Figure 4-8 show that the efficiency is closely dependent on the effectiveness of the heat exchanger. The curves of Figure 4-12 show that as the combustion temperature increases the optimum compression ratio for a given percentage heat recovery also increases. Thus at a given exchanger effectiveness of 90%, the optimum compression ratios are 3.5, 4, 6, 7.5, and 10 for combustion temperatures of 700°C, 770°C, 980°C, 1205°C and 2600°C, respectively. Contemporary turbines operate on combustion temperatures of 980°C.

Also, Figure 4-13 shows that one of the big advantages of heat recovery is a minor reduction in efficiency as the load decreases, which is not the case with a simple-cycle turbine.

Combined Gas/Steam Cycles

The maximum theoretical energy recovery from combustion requires the use of several complementary cycles. Thus the combination of a gas turbine and a steam turbine cycle in a closed circuit affords efficiencies higher than those possible with either of these cycles operating alone. This combined cycle permits increasing the power available with the gas turbine alone by 30% to 50%, depending on the turbine's efficiency. It consists of recovering part of the exhaust gas heat in a steam heater, of expanding the steam in a turbine connected to a driven machine whose

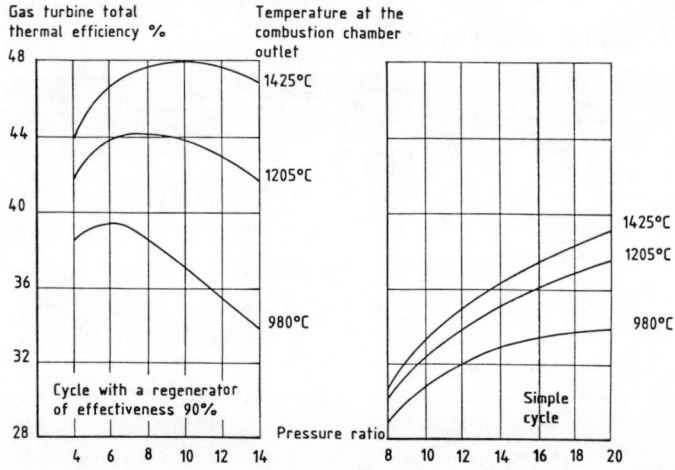

Figure 4-12. Influence of the combustion temperature on the optimum pressure ratio for a given percent heat recovery. (Courtesy of Garrett Corporation.)

shaft has two power outlets, and of condensing the expanded steam against air or water, the water condenser being less expensive and taking up less room. The steam turbine can be disconnected from the driven machine by a free-wheel type of automatic clutch that does not accept its load as long as the speed of the steam turbine is lower than the speed of the gas turbine. The steam heater can also be bypassed, at least if it is not a once-through steam generator.

The added investment for this system is large, as much as 2–3 times that of a gas turbine for the equivalent power. Control and regulation, which is entirely automatic, is on the amount of fuel gas to the gas turbine, with a set mount controller. The efficiency of the steam cycle is less than 25%.

The combined cycle can also be composed of a steam cycle common to several gas turbines, with one heater per turbine and one steam turbine driving an alternator or centrifugal compressor in parallel or series with the compressors driven by the gas turbines. This system has the following advantages:

- Lower investment costs.
- A steam turbine that costs less than several steam turbines with the same total power.
- Mechanical simplicity.
- The easiest method for adapting one or more existing gas turbines to a combined cycle.

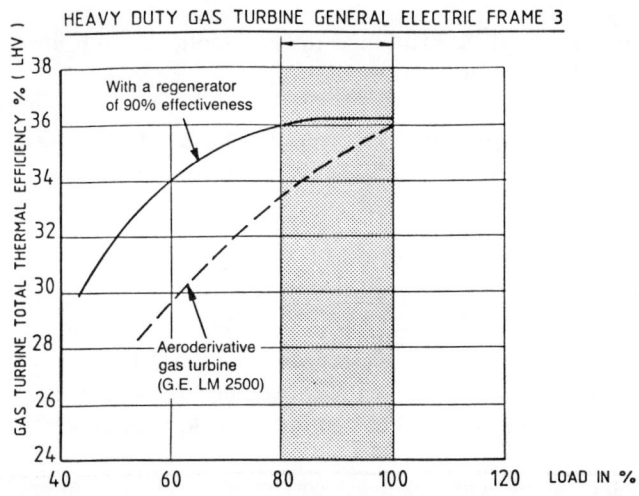

Figure 4-13. Comparing efficiencies at low loads for gas turbines with and without exhaust heat recovery. (Courtesy of General Electric.)

- No need for two shaft outlets on the compressors.
- Possibility for a high-efficiency compressor with axial inlet.
- A larger more efficient steam turbine.

When the compressors are in parallel, this combined cycle has the disadvantage of requiring use of all the gas turbines for the centrifugal compressor driven by the steam turbine to be able to operate. Indeed, if, for example, there are two gas turbines for one steam turbine and only one gas turbine in use, the power available on the steam cycle is reduced by 50%, and the compressor driven by the steam turbine cannot produce the same discharge head as the gas-turbine-driven compressor, which operates at 100% power. To overcome this disadvantage, the steam turbine must drive two compressors with one disconnected through clutching whenever one of the two gas turbines is shut down.

When the compressors operate in series, this combined cycle presents no disadvantage. It is best to put the steam-turbine-driven compressor last. Of course, the loss of one gas turbine brings on a reduction in the power of the steam turbine, and the corresponding reduction in total discharge head is greater than that corresponding to the loss of the gas turbine. This system is currently used for production of electric power where it has the advantage of conventional alternators with one shaft outlet. Some cycles possible with this system are discussed in Chapter 6.

Energy Balances with a Steam Cycle

Two cases will be examined under identical conditions of 12.5 mbar pressure drop in the suction filter:

1. A heavy duty type of turbine with average efficiency.
2. A turbine based on aircraft designs with high efficiency (aeroderivative).

For the first case of a heavy duty turbine, refer to Figure 4-14. The efficiency (LHV) of the gas turbine alone is 27%; the gross efficiency (LHV) of the combined cycle is 41%; and the net efficiency (LHV) of the combined cycle is 40.4%, with allowance for the pumps and air coolers. The increase in power of the combined cycle relative to the gas turbine is 52%.

For the second case of the aeroderivative gas turbine, refer to Figure 4-15. The efficiency of the gas turbine alone is 36.4%; the gross efficiency is 47.2%; and the net efficiency of the combined cycle is 46.7%.

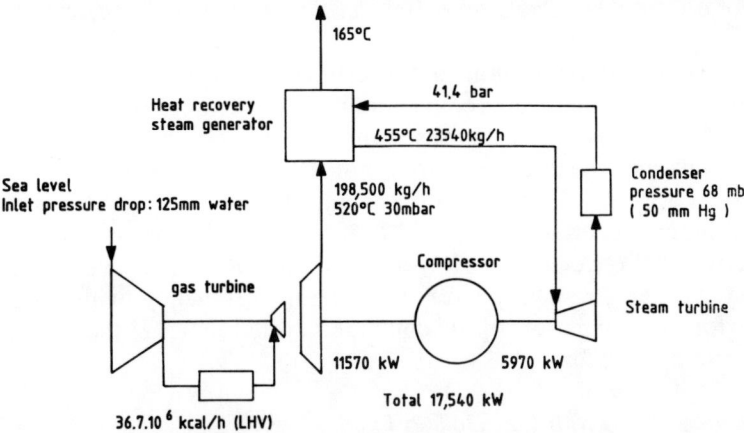

Figure 4-14. Heavy duty gas turbine flow scheme. (Courtesy of General Electric.)

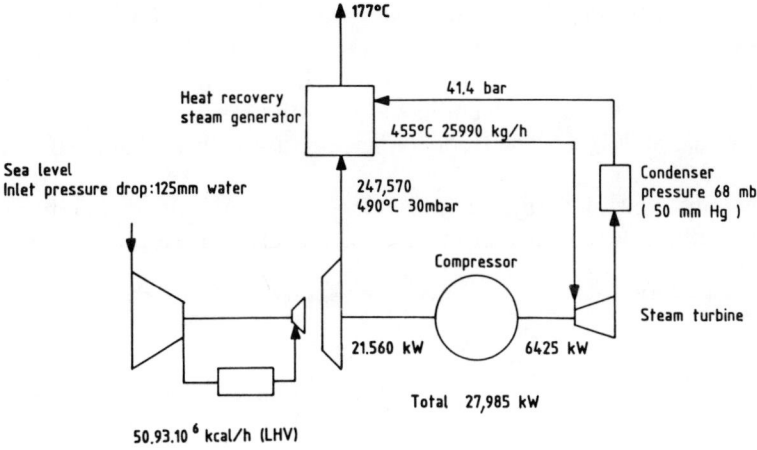

Figure 4-15. Aircraft-type gas turbine flow scheme. (Courtesy of General Electric.)

The increase in power of the combined cycle relative to the gas turbine is 30%.

The gas turbine with the lowest efficiency leads to relatively greater increases in power (52% versus 30%) because the energy available in the exhaust gases is relatively more. The gas turbine with the highest efficiency leads to a better combined-cycle efficiency, because the steam cycle has a markedly lower efficiency than the gas turbine and because the amount of energy recovered by the steam cycle is relatively lower for a less efficient gas turbine.

Use of a Combined Cycle on a Gas Turbine with Heat Recovery

It is possible to incorporate a gas turbine with heat recovery in a combined cycle, but the heat recovery unit adds little because it reduces the temperature of the exhaust gases going to steam generation and does not allow high enough steam pressure for an efficient steam cycle. Nevertheless such a cycle was put into operation in 1966 in Louisiana with the operating characteristics shown in Figure 4-16. The turbine would have had a thermal efficiency (LHV) of 21% without heat recovery; its efficiency with heat recovery is 28.5%. The net efficiency, with deduction made for the auxiliaries, of the combined cycle is 35.5%; the increase in power is only 25%.

Combined Cycle with Low-Boiling Fluid

A combined cycle with freon R 11 (trichlorofluoromethane, CCl_3F) was operated with a 865 kW Solar Saturn turbine in 1964, in order to use the low temperature of the exhaust gases and eliminate the need for make-up water (Figure 4-17). Freon R 11 boils at 15°C and 760 mmHg is practically nontoxic and nonflammable; its critical point is 198°C and 43.74 bar; and its rate of decomposition in the presence of steel is only 2% per year at 200°C. The freon expansion turbine has an adiabatic efficiency of 80%.

The only control of the freon cycle is for constant temperature of 150°C at the exit of the heater by opening and closing the control valve A

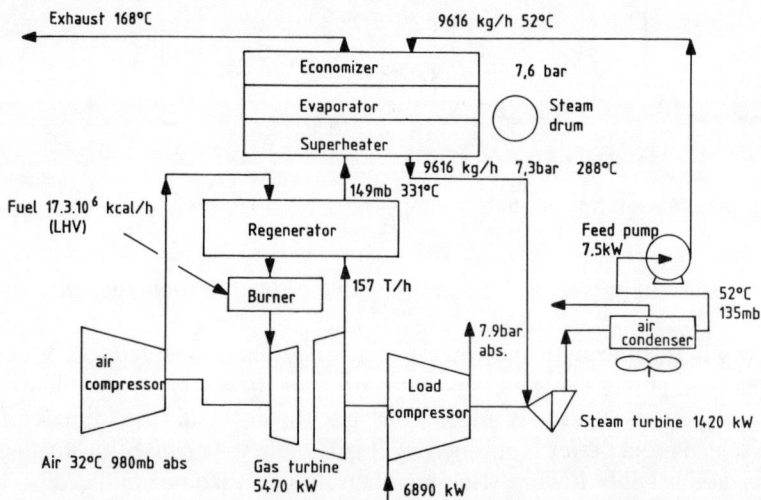

Figure 4-16. A gas turbine with waste heat recovery in a combined cycle.

Gas Turbines 137

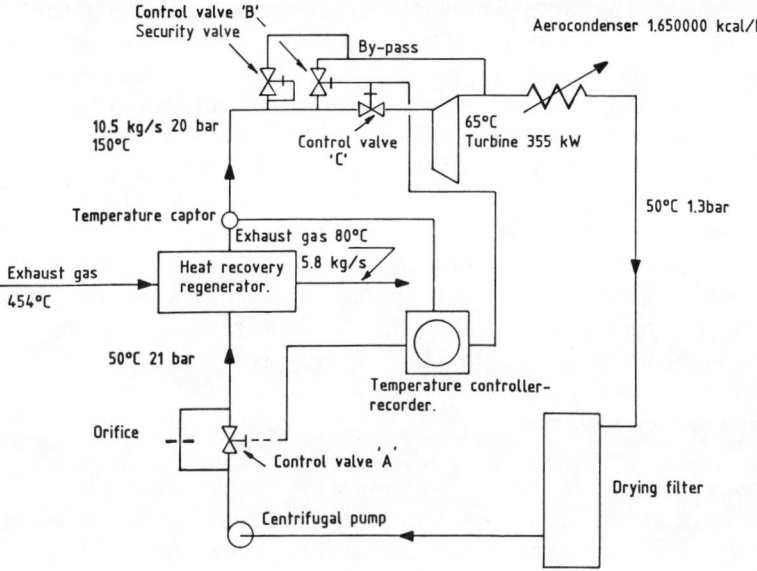

Figure 4-17. A combined-cycle gas turbine using Freon R 11 instead of water.

at the discharge of the feed pump. This regulator also controls inlet valve C and the freon bypass of the turbine, valve B (Figure 4-17).

Until the system is brought up to temperature, inlet valve C remains closed and bypass valve B is open. When operating temperature is reached, valve C is opened as valve B is closed. Because these two parallel valves are in series with valve A, their combined flow is proportional to the flow through valve A.

This combined cycle increases the turbine's power 42%; its thermal efficiency (LHV) increases from 22.3% to 31.6%.

Designing a Centrifugal Compressor Driven by a Gas Turbine Plus a Steam Turbine

The compressor must be drive-through and therefore have two bearings. High-efficiency compressors with one impeller and axial inlets cannot be used.

When more than one unit is operated the centrifugal compressors can be installed to run in parallel or to run in series.

If they are operated in parallel they must be designed to work at the desired head for the different conditions of ambient temperature whether they are driven by the gas turbine and steam turbine together, or by the gas turbine alone during startup periods or steam turbine shutdown for

138 Drivers for Rotating Equipment

maintenance. Therefore the operating range at nominal head must be the widest possible. This means that: the maximum compressor speed, which is 105% of nominal, should correspond to 100% nominal speed on the gas turbine, and the compressor's characteristic curves must be flat for greater stability. In winter, the operating point with the combined cycle should correspond to 105% of the compressor's nominal speed (Figure 4-18). The operating point under average conditions with the gas turbine alone must be clearly outside the surge zone.

If the compressors are operated in series, no particular requirements are imposed in the design of the compressors since when one is driven by the gas turbine alone, the total head decreases according to the throughput, following the pipe line characteristic curve.

Choice of Steam Turbine

Steam turbines are flexible within a wide range of speeds and power, and the steam turbine will not limit the combined cycle. However its nominal speed should be chosen, if possible, so that there is no gear between the turbine and compressor. The adiabatic efficiency of a steam turbine is 75%–80%.

Instrumentation and Control of the Combined Cycle

The auxiliary power consumers of the steam cycle are electric motors for pumps and air-cooler fans. Like the auxiliaries of the gas turbine, the level, flow rate, and temperature of the motors are controlled automatically. The only manual operation consists of purging and adding chemicals to the steam cycle.

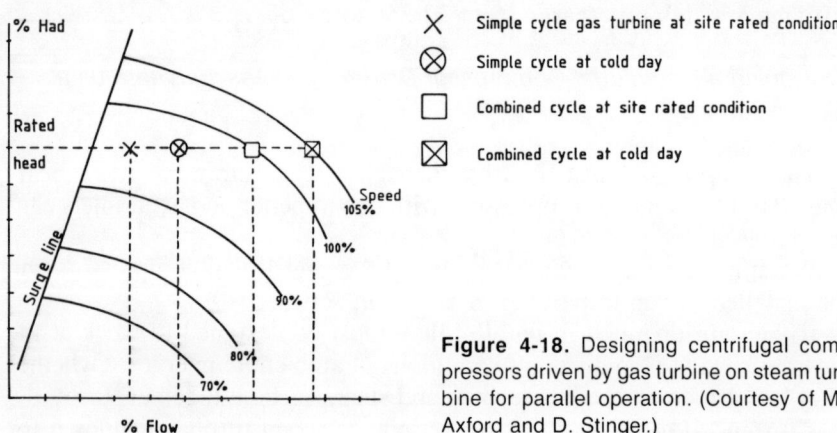

Figure 4-18. Designing centrifugal compressors driven by gas turbine on steam turbine for parallel operation. (Courtesy of M. Axford and D. Stinger.)

During ordinary operation, only the fuel flow to the gas turbine is controlled by a regulator, in the same way as for a gas turbine operating alone. The steam control valve at the inlet to the steam turbine remains wide open; and the turbine uses all available steam generated by the turbine's exhaust gases, so that the steam control valve acts as a shutdown valve when it simultaneously opens the bypass and sends the steam to the condenser (Figure 4-16).

In order to start up the combined cycle, the auxiliaries must be started and the levels regulated in the boiler drum and air condenser hot well (Figure 4-16). The gas turbine is purged and started; fuel is admitted, lit, and the turbine brought up to its nominal speed, which is maintained as the turbine is brought up to temperature, when the steam generator reaches its operating temperature and produces steam. This steam is first used to bring the piping up to temperature and to operate the seals of the steam turbine; then it is admitted to the steam turbine, which turns over and increases speed up to the speed of the gas turbine, at which time the steam turbine assumes part of the load of driving the compressor. After this period of coming up to temperature and starting up the steam turbine, the gas turbine comes under control of the regulator on its delivery. The startup takes about 30 minutes.

When operating conditions are changed, the steam cycle does not immediately follow the gas turbine cycle, but takes 2–3 minutes to become stabilized at a new equilibrium.

The combined cycles can be stopped merely by reducing the fuel gas flow to the gas turbine. After a short period of declining temperature, the fuel is shut off, and the inlet valve to the steam turbine can be closed by opening the bypass to send the steam directly to the condenser with an injection of water to remove superheat. In an emergency the inlet valve to the steam turbine can be closed and the steam either sent to the atmosphere or the condenser through the bypass valve or a safety valve (Figure 4-17).

Operation at Partial Load

When the gas turbine's controller reduces fuel flow and the exhaust gas temperature goes down, the flow, temperature, and pressure of the steam also decline, although the inlet valve remains wide open. The power of the steam turbine and efficiency of the steam cycle also is reduced. When the speed varies, the efficiency of the combined cycle, as well as that of the gas turbine alone, varies.

For example, the thermal efficiency on a curve corresponding to a centrifugal compressor where the load varies as (speed)$^{1/3}$ varies with the speed as follows:

Speed (%)	Thermal Efficiency (% of maximum efficiency)	
	Gas Turbine Alone	Combined Cycle
100	100	100
88.4	94	93
80.4	88	87

When the load varies at constant speed, the efficiency of the combined cycle, as well as that of the gas turbine alone, varies. Thus for load changes at a speed of 100%, thermal efficiency varies as follows:

Load (%)	Thermal Efficiency (% of maximum efficiency)	
	Gas Turbine Alone	Combined Cycle
100	100	100
69	93	94
52	86	87

Variations in ambient temperature do not significantly affect the efficiency of the combined cycle, because of compensation between the gas turbine and the steam turbine. When the ambient temperature increases, the temperature of the exhaust gases increases. Thus, the steam production, its temperature and pressure, and consequently the power produced by the steam turbine (despite the pressure of the air-cooled steam condenser), increases whereas the power produced by the gas turbine decreases. The increase in the steam turbine power compensates the reduction in efficiency of the gas turbine. With a water-cooled steam condenser, the condenser pressure can be kept constant and the efficiency of the combined cycle will increase slightly with the ambient temperature.

Designing Combined Cycles

Two types of boiler are used for combined cycles (Figures 4-19 and 4-20). One type, with heat exchange tube bundles and steam drums, must be bypassed when the steam cycle is not used. In this type the gas turbine has a conventional exhaust stack upstream of the boiler, with a valve system for sending the exhaust gases either to the stack or to the boiler. An example of this type (Figure 4-20) includes an Allison 570K gas turbine capable of 4800 kW at a thermal efficiency of 30% under ISO conditions. The combined cycle is capable of 6610 kW at an efficiency of 41.3%. The advantage of this gas turbine is in having the cold-side outlet opposite the exhaust, which greatly facilitates installation of the heater. On the

Gas Turbines 141

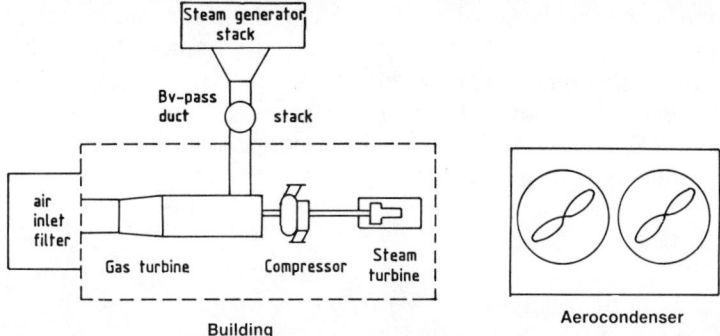

Figure 4-19. A combined-cycle gas turbine with the exhaust gas duct at right angles to the shaft line. (Courtesy of General Electric.)

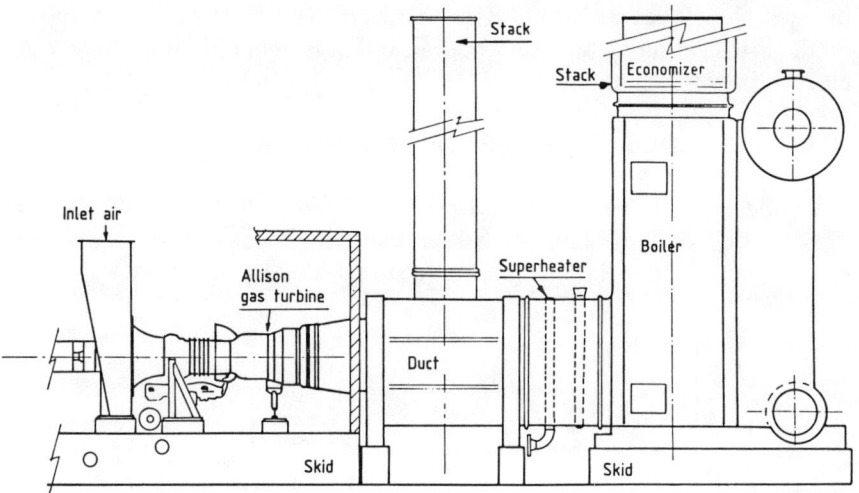

Figure 4-20. A combined-cycle gas turbine with inline exhaust duct and heater. (Courtesy of M. Axford and D. Stinger, Stewart and Stevenson.)

other hand, this type of turbine with two shafts has the disadvantage of having one drive shaft that crosses the entire machine, with a very inaccessible bearing in its center, that runs the risk of not turning if the speed of the gas generator equals the speed of the power turbine. In this case the ball bearing may be scored by axial thrust and stop turning correctly. The coupling between the compressor and steam turbine should be long enough for the compressor to be dismounted easily.

The big disadvantage of a boiler with a steam drum is that steam drums must be constantly attended, according to the laws of most countries. It is

142 Drivers for Rotating Equipment

this requirement that has held back the use of the combined cycle in pipeline compressor stations, which are normally remote-controlled and without night watchmen. This disadvantage can be overcome by a once-through steam generator that can stand exposure to the exhaust without cooling, in which case the bypass stack is not necessary except for maintenance shutdowns of the heater.

Trends in Combined Cycles

Increases in the compression ratios and temperatures possible with a gas turbine have brought constant improvement in the efficiency of the combined cycle, and contemporary conditions typically include compression ratios of 12–20, turbine inlet temperatures of 1090°–1370°C, and thermal efficiencies of 45%–49% (HHV). The maximum efficiency for the steam cycle levels off at 38% (Figure 4-21). (Efficiencies are based on the low heating value (HHV) because that is generally used for rating electric power generators.)

DESIGN CRITERIA FOR GAS TURBINES

The designs of specific gas turbines are generally based on whether there is one or two shafts, whether construction is aeroderivative or heavy, and whether the machine is first or second generation.

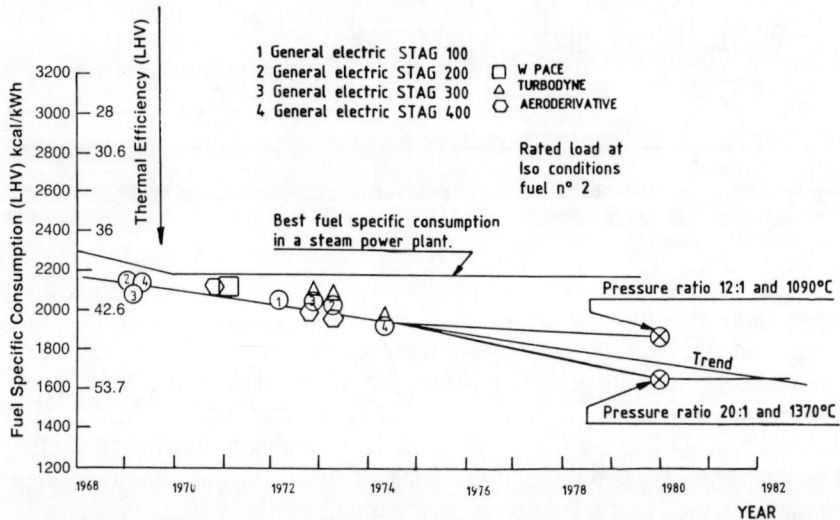

Figure 4-21. Trends in fuel efficiencies for the steam cycle of combined-cycle gas turbines. (Courtesy of General Electric.)

Choosing One or Two Shaft Lines

The shaft of single-shaft gas turbines is connected to (Figure 4-22) the air compressor, the expansion turbine, and the driven machine. The expansion turbine furnishes the power both to drive the air compressor and to supply the user. However, these two functions can be dissociated, so that the gas turbine has two shaft lines (Figure 4-23).

When there are two shafts, the first comprises the air compressor and a high pressure turbine sized to meet the air compressor's power needs; this combination is called the "gas generator." The second shaft line comprises the power or low-pressure turbine and the driven equipment. Although more complex, this turbine with two shaft lines is justified for gas transmission because of its greater flexibility. The two shaft lines can each operate at different speeds of rotation and thus offer a wider range of speeds for the driven machine, as well as facilitate startup with the driven machine automatically uncoupled. Because it is independent of the speed of the gas generator, changes in the speed of the power turbine do not act on the available power, which can remain close to nominal. Thus

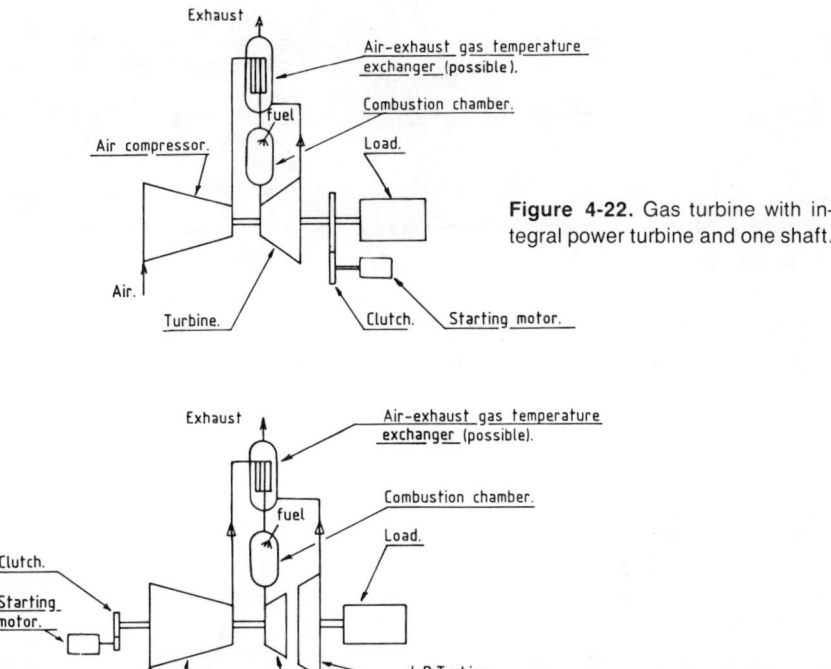

Figure 4-22. Gas turbine with integral power turbine and one shaft.

Figure 4-23. Gas turbine with free power turbine and two shafts.

144 Drivers for Rotating Equipment

the possible variations in speed of the power turbine can be rather wide without affecting the efficiency too much. This means higher efficiency at partial loads. Startups and shutdowns are faster. And in the case of a breakdown, there is a possibility for quickly changing the gas generator when the construction corresponds to a power turbine placed at the outlet of a jet engine more or less modified for industrial use (Figure 4-24).

Construction with a single shaft is suited for machines, particularly alternators, whose speeds are nearly constant, whereas the construction with a two-shaft machine is well suited for driving compressors and pumps. Machines with two shafts permit speed variations 20%–30% wider than machines with one shaft (Figure 4-25). Power is limited by the inlet temperature to the first impeller of the expansion turbine, and for turbines with two shafts by the rotating speed of the gas generator.

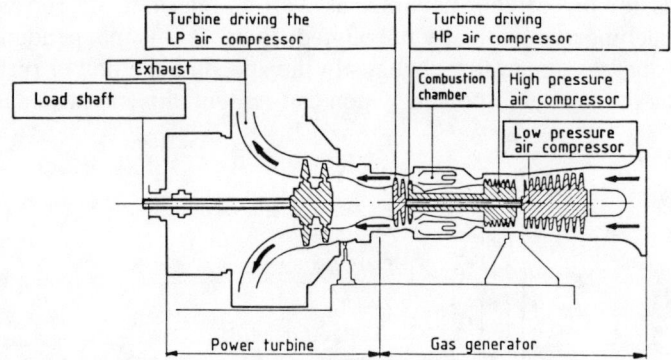

Figure 4-24. Typical two-shafts simple-cycle aeroderivative two-spools gas turbine.

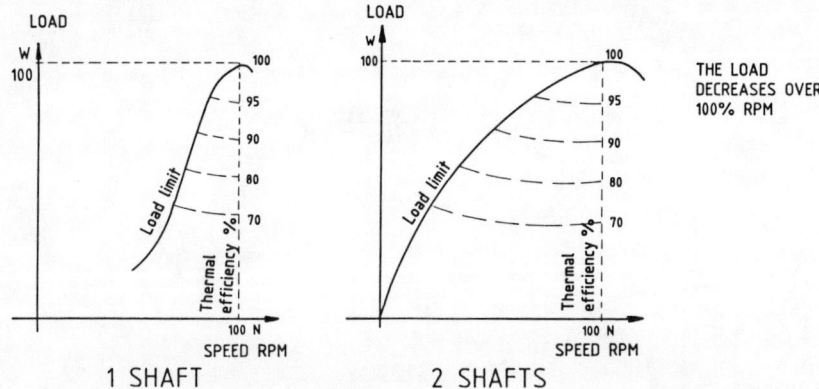

Figure 4-25. Two-shafts gas turbines have broader speed characteristics.

Light and Heavy Turbines

Some gas turbines (called heavy duty turbines) are designed specifically for industrial application; they are based on conventional steam turbine technology and high power, very often for the production of electricity. These applications often require heavy construction and large sizes. Such turbines, which depend on fixed installations, are calculated for long endurance and a high coefficient of on-stream time. Their power can go over 75 MW, particularly when they are combined with the production of steam (Figures 4-26 and 4-27).

By contrast, aeronautical design and fabrication technology has produced light- and medium-weight turbines with several important advantages that suit them for industrial applications. They are reliable; hot parts last for over 40,000 hours with general review required only every

Figure 4-26. Two-shafts heavy duty gas turbine. (Courtesy of General Electric.)

Figure 4-27. Sulzer 10,550 kW heavy duty gas turbine. (Courtesy of Sulzer.)

146 Drivers for Rotating Equipment

30,000 hours (3+ years). They are convenient; it is easy to perform maintenance on the site without disaligning the machine; and parts are readily accessible for changing or cleaning. They can operate with various gas and liquid fuels. They are relatively compact and light, permitting delivery in prefabricated groups (Figures 4-28, 4-29, 4-30, and 4-31) and fast and easy installation. They can withstand frequent startup and very rapidly assume loads in as little as 30 seconds to 2 minutes.

First- and Second-Generation Turbines

Open-cycle turbines are classified as first or second generation.

Second-generation turbines have an efficiency over 29%, for machines of 3,000 kW, and over 35% for machines of more than 15,000 kW, which later corresponds to a compression ratio of 12 and a combustion temperature of more than 1000°C. The thermal efficiency decreases less with increasing loads and speed. The curve of power vs. speed is much flatter, as shown in Figure 4-32.

Manufacturers of aviation turbines adapt their products to industrial applications by using their turboreactors as direct gas generators by means of a few modifications, such as replacing ball bearings with jour-

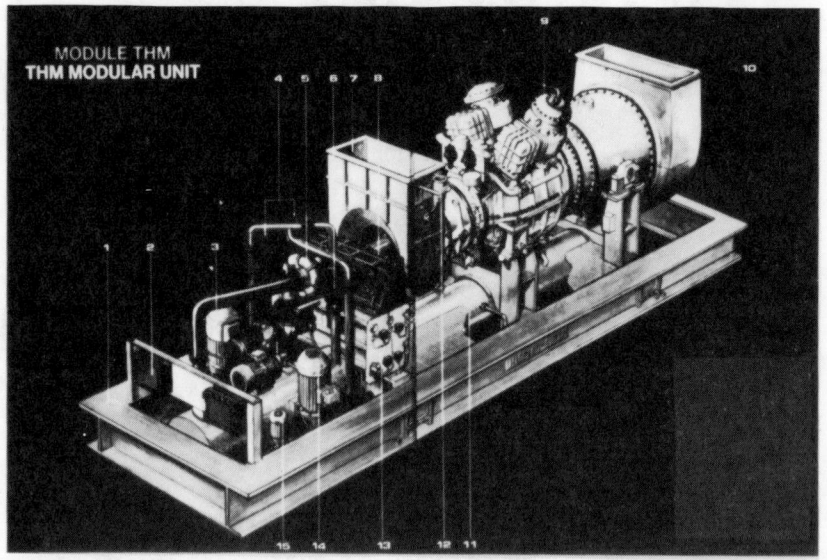

1. Sub base
2. Electrical junction box
3. Stand-by motor driven LO pump
4. Lube oil filters
5. Expander starting system
6. Main LO pump (turbine driven)
7. Air intake elbow
8. Auxiliary gearbox
9. Combustion chamber
10. Exhaust elbow
11. Lube oil tank
12. Compressor cleaning system
13. LO system gauge panel
14. Emergency LO pump (motor driven)
15. Electric circulating pump for LO heating system

Figure 4-28. THM Hispano gas turbine. (Courtesy of Hispano Suiza.)

Gas Turbines 147

Figure 4-29. Solar Mars gas turbine. (Courtesy of Solar Gas Turbines, Inc.)

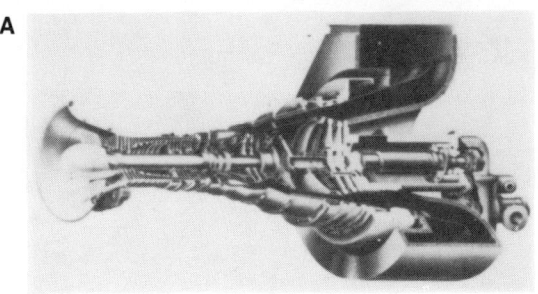

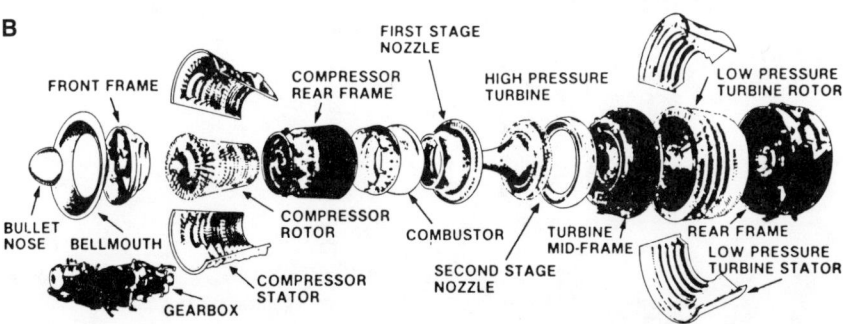

Figure 4-30. (A) An Ingersoll Rand power turbine; (B) with a matching General Electric gas generator.

148 Drivers for Rotating Equipment

Figure 4-31. Packaged 10,500 BHP Cooper Bessemer gas-turbine-driven compressor station (foreground). The compressor station in the background has the same installed power, using reciprocating engines.

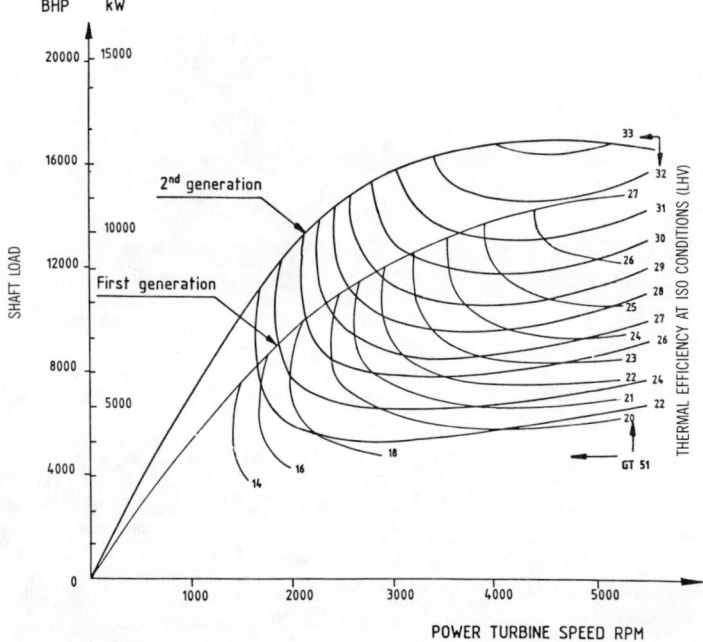

Figure 4-32. Constant-efficiency lines and characteristic curves for first- and second-generation gas turbines. (Courtesy of Ingersoll Rand.)

nal bearings and eliminating the turbofan. The resulting gas generator is then combined with a power turbine fabricated by either the manufacturer of the gas generators or by another company to whom they are sold (Figure 4-30). The reduction in the loads and peak temperatures reached during jet-plane take-off, as well as the sudden variations in operating conditions due to changing altitude permits durations of life comparable to heavy- or medium-weight turbines with reviews spaced at 30,000 hours. Although the manufacturing technology for these high-efficiency machines is complex, the gas generator's low weight and ease of dismounting make it easy to return it to the factory for periodic checkup.

The gas generator can assume its load in a few seconds after starting, because it has a low thermal inertia; however the power turbine is designed as a medium-weight turbine with one or two overhung impellers, and its thermal mass is therefore large due to the large expansion per impeller, so that the turbine requires a warm-up period of 5–20 minutes depending on the machine.

GAS TURBINE DESIGN

Gas turbine design can be approached through each of its functional parts taken separately, i.e., through the air compressor, the combustion chamber, and the expansion turbine, plus the sometimes exotic materials of construction.

Air Compressor

Since gas turbines were developed from either steam turbines or aviation turboreactors with a small volume for a high power, the air compressor is usually axial, although a centrifugal compressor has definite advantages. The axial compressor has a straight characteristic curve and thus an inflexible operation with a relatively small stability zone; it will enter into surge as soon as the rotating speed is low enough to induce an air flow too low for the back pressure put up by the expansion turbine, where the flow at the inlet is limited by the speed of sound. This is the case during startup and shutdown. Either one or more of the antisurge valves, adjustable stator vanes, or multiple rotors are used to overcome this problem.

Opening antisurge bleed valves increases the flow in the first compression stages during compression.

Adjustable stator vanes on the first compressor stages can have their position controlled as a function of the rotating speed. This system is highly useful for two-shaft turbines where the speed of the gas generator varies with the power. The adjustable vanes improve efficiency at partial load by adapting the speed triangles. It should be emphasized that gas turbines have the disadvantage of efficiencies that decline rapidly with decreasing load.

Two compressor rotors (or even three in aviation turboreactors equipped with a turbofan) with each rotor driven by one or two impellers of the expansion turbine, enable the rotor speeds to be adjusted automatically as a function of the required load. This solution is obviously possible only with turbines with two shafts. The result is similar to adjusting the vanes with respect to improving the efficiency at partial load.

The highest possible compressor efficiency requires minimum clearance for minimum leakage between rotor and stator. To achieve the minimum clearance in high-efficiency machines, a layer of weld metal or alumina, or even a ring of synthetic rubber loaded with metal, is spray-deposited at the vane passages on the stator and rotor, and these coatings are worn smooth during break-in of the machine.

Both fixed and moving vanes are treated to have a better surface. Treatments may be:

- A 15-micron layer of nickel-cadmium (Hispano Suiza). This has the disadvantages of a more rapid fouling of the surface and subsequent difficulty in recovering the surface characteristics after cleaning, as well as fragility, which forestalls cleaning the compressor with abricotine and permits washing only with aqueous solvent solution.
- An aluminum paint mixed with an inorganic binder and baked at 300°C (Sermetel 53-75). The surface hardness of this coating is double that of the nickel-cadmium layer, and the surface is better. This product is used on fixed vanes by General Electric. The treatment on the moving vanes is nickel cadmium on the discs, or a very hard ceramic that is very smooth (H3-51).

The air compressor can also be a centrifugal compressor with one or two stages, and the axial compressor can also have a final centrifugal stage. However, centrifugal air compressors suit only smaller machines with less than 4,000 kW, because centrifugal compressors are much larger than axial ones for the same air flow. On the other hand, centrifugal compressors have the important advantages of stability, flexibility in speed, and resistance to fouling and foreign matter. Their stability makes antisurge valves unnecessary; because of their speed flexibility, their efficiency falls off less at partial load; and their resistance to fouling enables maintaining the compression ratio without frequent cleaning.

Combustion Chamber

Combustion chambers (Figure 4-33) can be either completely separated from the turbine or placed around the turbine in barrel or annular

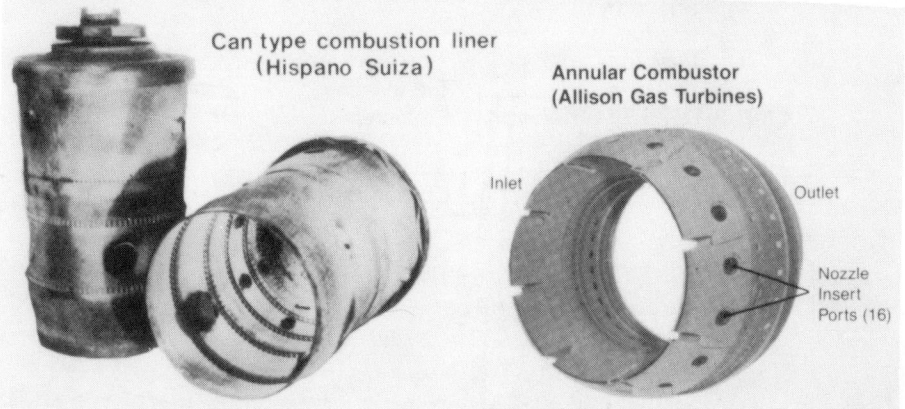

Figure 4-33. Two versions of combustion chamber liner.

form. The separated or placed in barrel chambers can be largely dimensioned and are easily accessible. The annular chambers are less accessible but the air path is less disturbed and the pressure drop is lower.

There will be an injector for liquid or gaseous fuel with an introduction of secondary air all along the chamber, plus external air cooling. Around 40% of the compressor air flow is used for cooling hot parts of the combustion chamber, expansion turbine and discs, and fixed and moving vanes of the expansion turbine.

Expansion Turbine

The expansion turbines are axial with several stages, except for small turbines under 500 kW, which may be centripetal. In two-shaft gas turbines the expansion turbine is divided in two with one part called the power or "free" turbine driving the receiving machine and the other part driving the gas generator. The power turbine, which includes 1-6 stages, is not exposed to high temperatures and does not represent a difficult service.

The turbine driving the gas generator can have 1-3 stages, of which the first stages are exposed to the highest temperatures; and its discs as well as fixed and moving vanes are cooled with circulating air on the inside and even a laminar film of air on the vanes' surfaces by means of holes (Figure 4-34). The cooling air is supplied by the air compressor.

In the General Electric heavy duty gas turbines the expansion turbine has directing vanes that can be adjusted to improve efficiency when the load and speed vary.

Materials for the vanes and discs are chosen as a function of the temperature to resist wear, corrosion, and creep (Table 4-3).

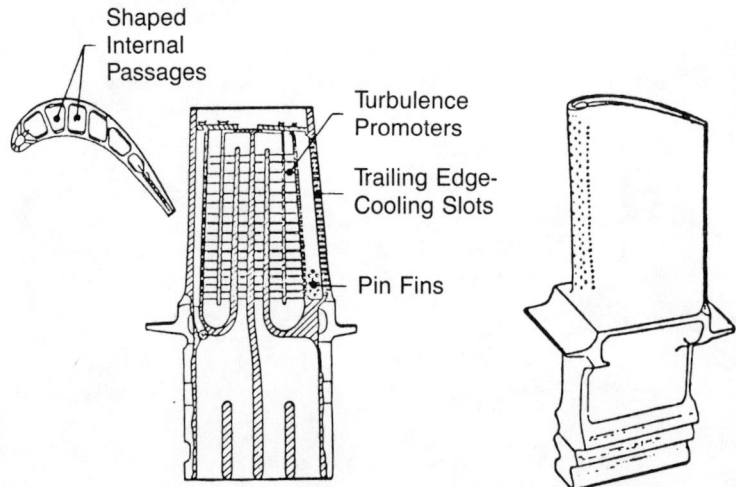

Figure 4-34. Construction of high-temperature gas turbine blades. (Courtesy of General Electric.)

Choice of Materials

The choice of materials is of great importance. Typical materials used in high efficiency (over 30%) turbines at high temperatures (over 1000°C) and high compression ratios (over 16) are shown in Table 4-1.

REDUCING NITROGEN OXIDES (NO$_x$) IN THE EXHAUST

Although laws in the United States and other countries limit the nitrogen oxides in the exhaust gases of turbines with more than 4,000 kW power, the technical means for meeting these limitations have yet to be perfected. The formation of NO$_x$ in the internal combustion engine has been described in Chapter 2.

Figure 2-9 shows that the NO$_x$ concentration in the exhaust gases increases with the temperatures of the combustion (which depends on the gases entering the high pressure turbine), with use of liquid rather than gaseous fuels, and with increasing pressure in the combustion chamber, and consequently the compression ratio. Thus the NO$_x$ concentration when using distillate fuels is 60%-90% higher than when using natural gas, and it will increase from about 185 ppmv for a compression ratio of 18 to about 267 ppmv for a compression ratio of 29.

It should be noted that the carbon monoxide concentration diminishes as the NO$_x$ concentration increases.

Research in reducing NO$_x$ emissions has evolved around injections to reduce the flame temperature, multistaged combustion, combustion

Table 4-3
Materials Used in Fabricating Gas-Turbine Parts Exposed to Heat

Type of Turbine	Solar Mars	Cooper Coberra 3045	Cooper Coberra 6156	Ingersoll GT 61	Ingersoll GT 71
Gas generator					
Compressor rotor:					
Moving vanes	17-4 PH	Titanium alloy	Titanium alloy	T1-6A1-4V A-286	Titanium alloy
Fixed vanes	17-4 PH	Corrosion resistant material	Corrosion resistant material	T1-6A1-4V A-286	Titanium alloy
Discs	—	"	"	T1-6A1-4V	—
Shaft	AISI 410	Chrome steel	Cr-steel/Inco 901	Inco 718	Inco 718
Rear end	—	Corrosion resistant material	Corrosion resistant material	Inco 718	—
Combustion chamber	Hastelloy X	Nimonic C 263/75	Nimonic C 263	Hastelloy X/Hs-188	Hastelloy X/Hs-188
Turbine bracing	—	Chrome steel	Chrome steel	Inco 78/Rene 41 Hastelloy X	—
Turbine rotor					
Moving vanes	Cobalt steel	Nickel-chrome steel	Nickel-chrome steel	X 40/Rene 80	X 40/Rene 80
Fixed vanes	Mar-M-421	Cobalt steel	Cobalt steel	Rene 80	Rene 80/77
Power turbine					
Moving vanes	Inco 713 C	Udimet 500/X-40	Udimet 500/X-40	Inco 718/Waspaloy	Udimet 500

(table continued on next page)

Table 4-3
Continued

Type of Turbine	Solar Mars	Cooper Coberra 3045	Cooper Coberra 6156	Ingersoll GT 61	Ingersoll GT 71
Fixed vanes	N-155	N-155	N-155	Udimet 500/N-155	N-155/AISI 347
Discs	A-286/V 57	A-286 AISI 4340	A-286 AISI 4340	AISI 4340	AISI 4340

Analysis of principal alloys used in gas turbines

Material	Ni	Co	Cy	Fe	Al	Ti	Mo	W	Cb	Ta	B	Zy
Rene 80	Balance	9.5	14.0	—	3.0	5.0	4.0	—	—	—	0.015	0.03
Udimet 500	Balance	18.5	18.5	—	3.0	3.0	4.0	—	—	—	0.010	—
Inco 713 C	Balance	—	12.5	—	6.1	0.8	4.1	—	2.0	—	0.012	0.01
Mar-M-421	Balance	10.0	15.5	—	4.25	1.75	1.75	3.5	1.75	—	0.015	0.05
X-40	10.0	Balance	25.0	1.0	—	—	—	7.5	—	—	—	—
Hastelloy X	Balance	1.5	22.5	18.5	—	—	9.0	0.6	—	—	—	—
Inconel 718	Balance	—	19.0	19.0	0.4	0.9	3.0	—	5.0	—	—	—
Waspaloy	Balance	13.5	20.0	—	1.2	2.9	4.3	—	1.0	—	0.006	0.08
A 286	26.0	—	15.0	Balance	0.2	2.0	1.3	—	—	—	0.015	—
N 155	20.0	20.0	21.0	Balance	—	—	3.0	2.5	—	—	—	0.15
Nimonic 263	Balance	19.0	19.5	0.3	0.5	2.2	5.9	—	—	—	—	—

within a vortex, premixed combustors, inhibiting NO_x-forming reactions, and catalytic reduction of NO_x.

Injections to Reduce Flame Temperatures

These injections have been of water, steam, or nitrogen. Water may be injected either into the burner nozzle (Figure 4-35) or at the burner outlet (Figure 4-36). In either case, the water must be fully demineralized to less than 0.5 ppm in weight of metal contaminants (Na, Li, K, Pb, Va) and less than 2.65 g/m³ solids. Thus it becomes expensive. The rate of water injections should be 0.5–0.6 units, the weight of fuel, in order to achieve an NO_x reduction of 70%–80%—70% for natural gas fuel and 80% with distillate fuels. Higher rates of injection achieve progressively less improvement (Figure 4-37). Also, water injection reduces the thermal efficiency as indicated in Figure 4-38.

As more and more water is added, a point is reached at which a sharp increase in carbon monoxide is observed. This point defines maximum water injection level for a given combustion system. The higher the turbine inlet temperature the more tolerant the combustion is to the addition of water.

Because the formation of NO_x varies with operating conditions, water injections can be used intermittently, rather than continuously, to maintain the NO_x emissions below a prescribed level, such as, for example, 150 ppmv. This can be done automatically; and it is feasible to have controllers operated by measurements of ambient temperaure, power output,

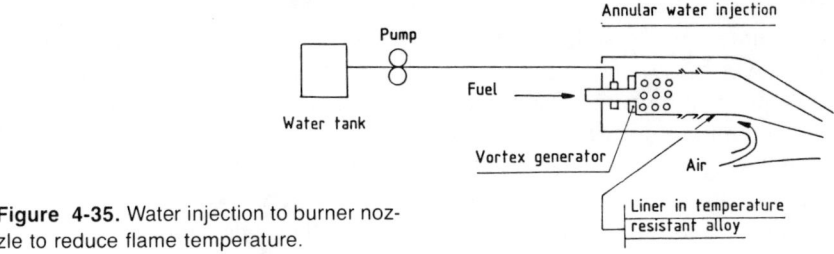

Figure 4-35. Water injection to burner nozzle to reduce flame temperature.

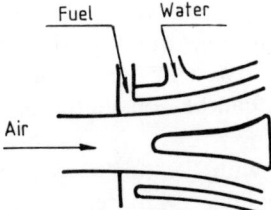

Figure 4-36. Schematic view of a burner with water injection at the outlet.

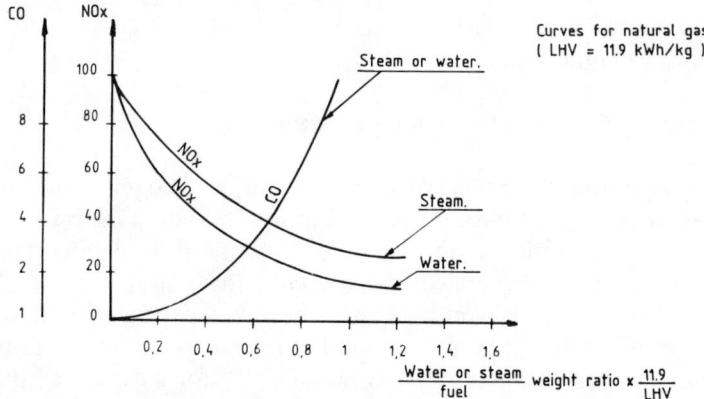

Figure 4-37. Influence of water and steam injections on pollutant concentration in exhaust gases.

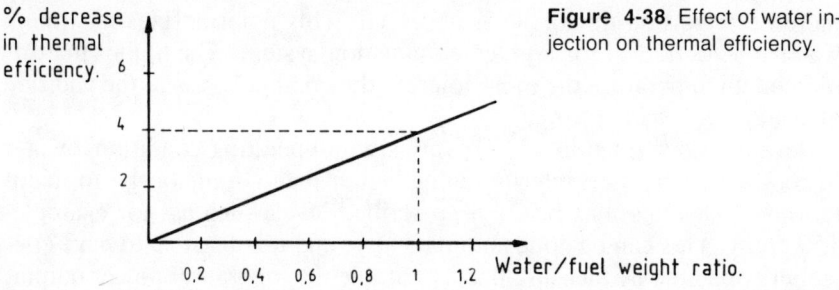

Figure 4-38. Effect of water injection on thermal efficiency.

and ambient air humidity to vary the water injection so as to maintain the NO_x level below one of the constant concentration lines shown in Figure 4-39. It must be noted that the carbon monoxide concentration increases as the temperature of combustion diminishes analogous to the curves of Figure 4-39.

Steam injections are mixed into natural gas fuel before the burner's nozzle or into liquid fuels at the burner's outlet. The steam is saturated, dry, and at the pressure of the air at the burner inlet. Both the power and efficiency are increased by these injections, when the steam is generated by a boiler in the turbine's exhaust. For example, a given turbine using water or steam injection for reducing NO_x emissions might have an efficiency of 33.5% with water injections and of 36% with steam injections. However, injections of either water or steam can be considered as only provisional because of their cost.

On the other hand, if steam is generated with the turbine's exhaust gases, it can be used to both reduce NO_x emissions and increase efficiencies. This is a parallel combination of the Brayton and Rankine cycles called a Cheng cycle. The amount of steam for this purpose is much larger than for NO_x reduction alone and will come to 7–14% by weight of the combustion air. This steam can be injected either through a distributing ring at the outlet of the air compressor, or into the burner, or between the high pressure and low pressure turbines through a distributing ring. The improvement in thermal efficiency obtained with a second generation turbine in this way will amount to 6–7 points of the existing efficiency and reaches 40% to 43%.

Finally, the injection of nitrogen will reduce NO_x emissions as shown in Figure 4-40. If this nitrogen occurs within a natural gas fuel, it is free; but if it is necessary to add it, it becomes too expensive.

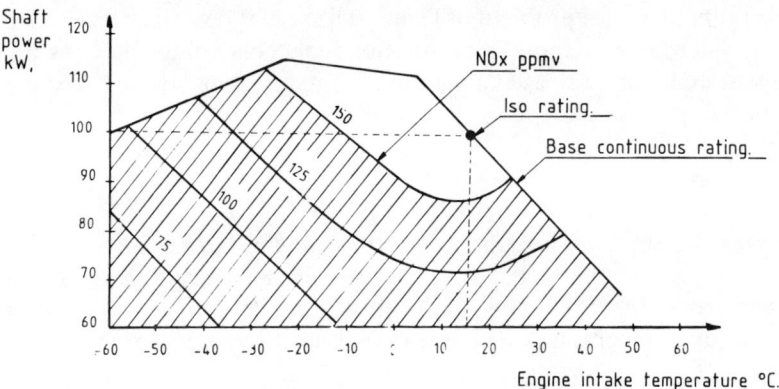

Figure 4-39. Operating diagram for automatic water injection.

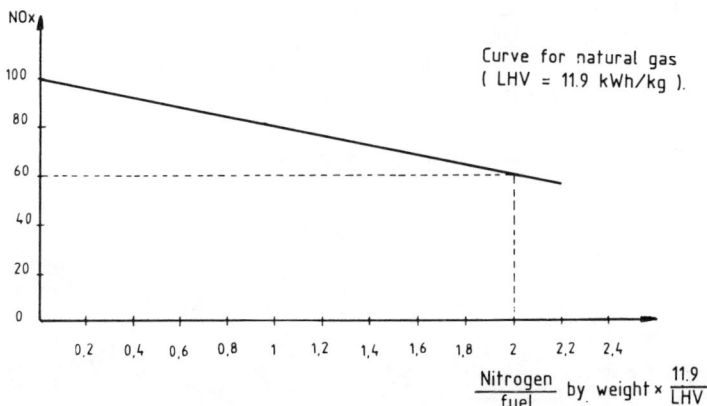

Figure 4-40. Influence of nitrogen injections on NO_x.

Multistaged Combustion

By introducing the fuel and combustion air at two points in the combustion chamber (Figure 4-41), it is possible to carry out combustion in lean mixtures that avoid the peak adiabatic temperatures. And this reduces the NO_x content of the exhaust. However control of this system is complex, and to have stable combustion it is necessary to have a catalyst such as platinum on a ceramic base placed in the combustion zone.

Combustion within an Induced Vortex

This system uses the laws of fluid dynamics to accomplish a staged combustion analogous to that of Figure 4-42. This system is thus much simpler than multistaged combustion with respect to control (Figure 4-43), but there is a problem of flame stability, as described in Chapter 2. This system is extremely attractive for fuels containing FBN, as the nitrogen atoms are released in a zone containing insufficient oxygen to form NO.

Lean Combustion

When the fuel-air mixture is obtained at the burner in conventional combustion chambers, the fuel-air ratio is normally 1 and cannot be lower than 0.71 for flame stability. It is possible to have stable combustion at low temperatures with a lean fuel-air mixture at a richness as low

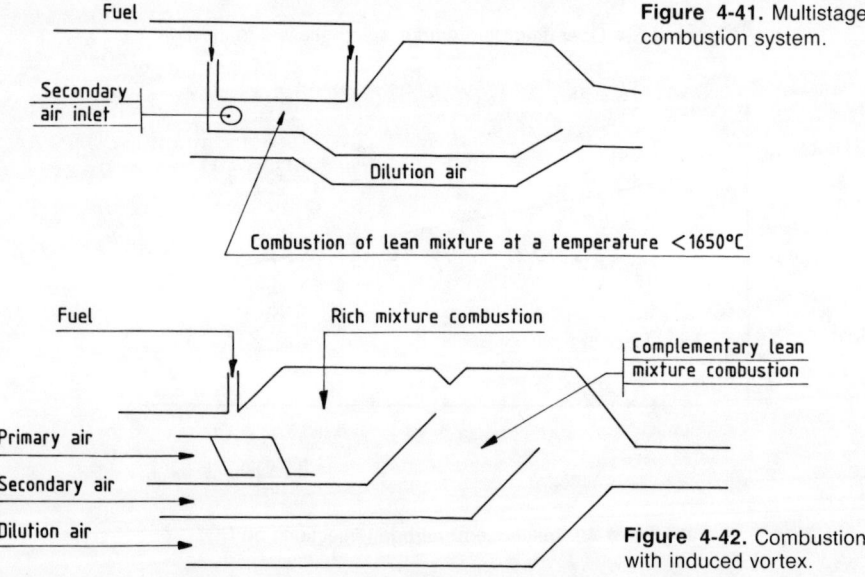

Figure 4-41. Multistage combustion system.

Figure 4-42. Combustion with induced vortex.

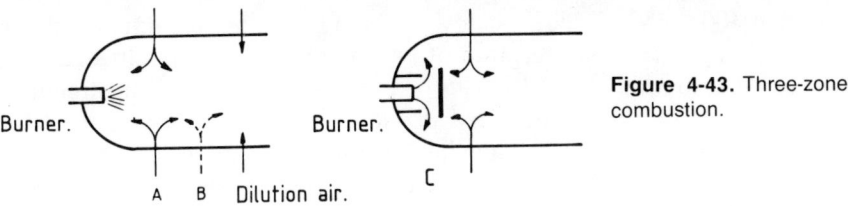

Figure 4-43. Three-zone combustion.

as 0.60 (air-fuel equivalence ratio of 1.66) if the mixture is perfectly homogeneous. Such homogeneity can be obtained with a swirl stabilized premixed combustion system. The NO_x content can be as low as 10 ppmv at 15% O_2.

Inhibiting NO_x-Forming Reactions

NO_x formation is inhibited by reducing the combustion temperature, which can be reduced through the sensible heat of the excess air used in gas turbines. Towards this end the combustion chamber can be considered as having three zones (Figure 4-43). In the first zone, the fuel finds its stoichiometric match of air of combustion, and a high temperature flame is created. In the second zone, additional air enters the mixture to make it leaner as the combustion continues. In the third zone, still more air dilutes the mixture of combustion products.

Experience has shown that, if the temperature of the second zone is reduced to 1700 K or less, the NO_x concentration in the exhaust can be reduced by 40%. If the temperature in the first zone is held to 2200 K and the second zone to 2000 K, as the duration of the latter is reduced, it should be possible to achieve a 70% reduction in NO_x concentrations. This may be accomplished through combustion chamber construction as shown in Figure 4-43. Two techniques are possible. The point of secondary air injection is on the left; nearer the flame with the more important rate: A instead of B. The flame and period of high temperature are reduced by increasing turbulence and recirculation around an obstacle before the burner, as shown on the right. These techniques resemble techniques currently being applied to so-called "high turbulence burners" used in modern industrial furnaces.

Catalytic Reaction of Nitrogen Oxides

It is possible to catalytically reduce the NO_x in the exhaust of a gas turbine through the injection of reducing ammonia (NH_3) upstream of a catalytic reactor. This technique also is possible with internal combustion engines. However, the metal catalyst used for this reaction deteriorates rapidly at temperatures higher than about 450°C; and the effective range is about 300°–450°C. The window shifts to higher temperatures with the catalyst's age. Thus the reaction of NO_x in the exhaust requires cooling the exhaust to around 500°C with possibly a boiler located upstream of the reactor. The steam generated in such a boiler can then be used for injection to the combustion chamber for improving the reaction process. However, it is difficult to continuously measure the concentration of NO_x in the exhaust gases, and thus to automatically control the flow of NH_3 reducing agent.

EXISTING GAS TURBINES

A wide range of gas turbines compete for industrial applications, including light turbines of 50–10,000 kW, heavy industrial turbines of 4,000–75,000 kW, and turbines with aeronautical gas generators that are capable of 3,000–24,000 kW. As an example of the latter, General Electric has developed a gas generator, No. LM 5000, with a combustion temperature of 1200°C and a compression ratio of 29, which when coupled with a power turbine provides a gas turbine capable of 35,000 kW at a thermal efficiency of 38%. This gas generator is derived from G.E.'s turboreactor CF6-50, which has been in use since 1972 on airplanes such as the Airbus A 300.

All gas turbines are now mounted on single or multiple skids. They comprise light, independent, compact groups needing only simple foundations such as a slab of reinforced concrete. They can be mounted in a building within sound proof walls or outside, within a sound proof enclosure. Their thermal efficiencies (LHV) are 18%–22% for less than 1,500 kW, 20%–25% for first generation turbines less than 10,000 kW, and over 30% for second generation turbines with compression ratios over 12 and temperatures over 1000°C. More specifically, these latter will have efficiencies of 30% for 4,000–6,000 kW, 33% for 6,000–12,000 kW, and 36% for more than 20,000 kW.

OPERATING CHARACTERISTICS AND USE

The operating ranges of gas turbines are defined through their characteristic performance curves, their exhaust temperatures, their operating ranges, their oil consumptions, and their startup requirements.

Characteristic Curves

The characteristic operating curves are given as curves of power limit vs. speed of the power turbine (Figures 4-44, 4-45, and 4-46) at the reference conditions. The power limits are plotted as parameters of constant mass flow or constant ambient air temperature, which are equivalent because the mass flow is inversely proportional to the suction air tempera-

Gas Turbines 161

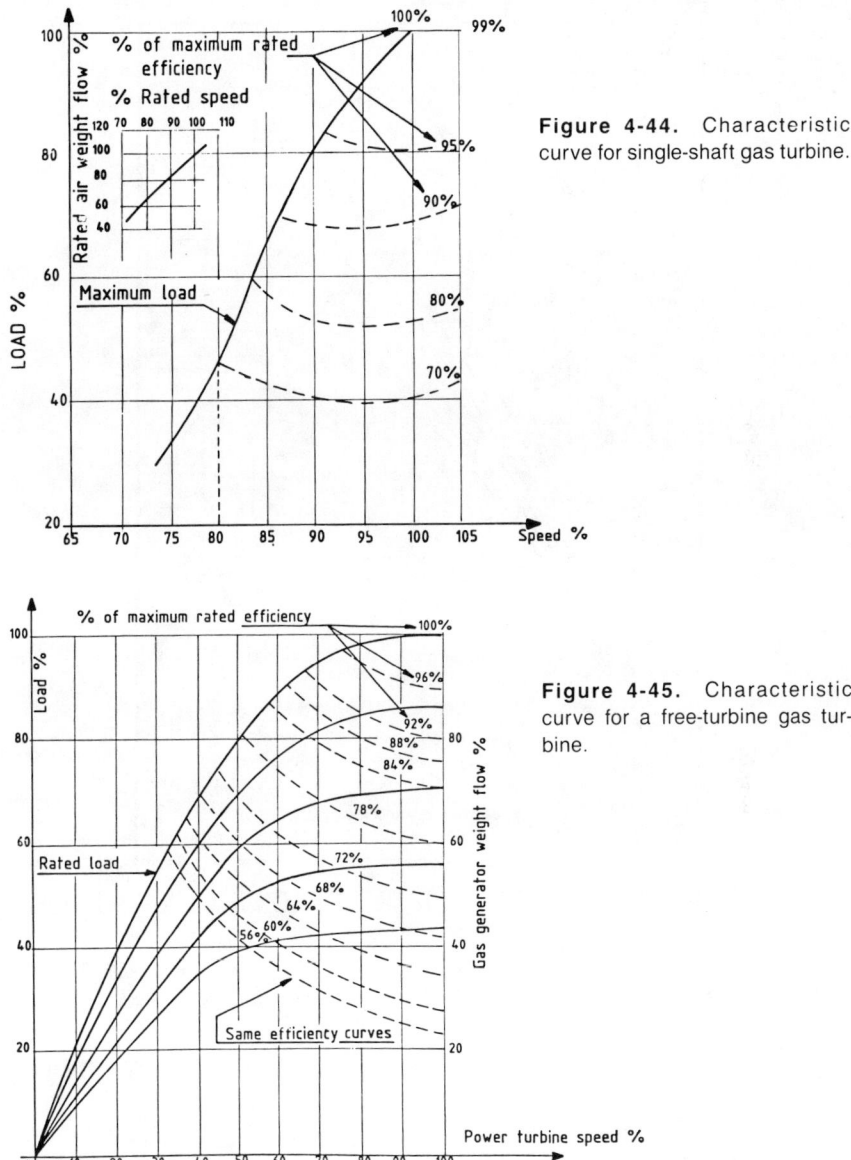

Figure 4-44. Characteristic curve for single-shaft gas turbine.

Figure 4-45. Characteristic curve for a free-turbine gas turbine.

ture. The mass flow parameters are usually shown as percents of the nominal 100% mass flow corresponding to 15°C for ISO conditions at sea level and 80°F for NEMA standard conditions at 1,000-foot altitudes. If the gas turbine has two shafts, the parameter lines may be in terms of the gas generator's rotating speed, which corresponds approximately to mass flow curves when the air compressor is axial, because its

162 Drivers for Rotating Equipment

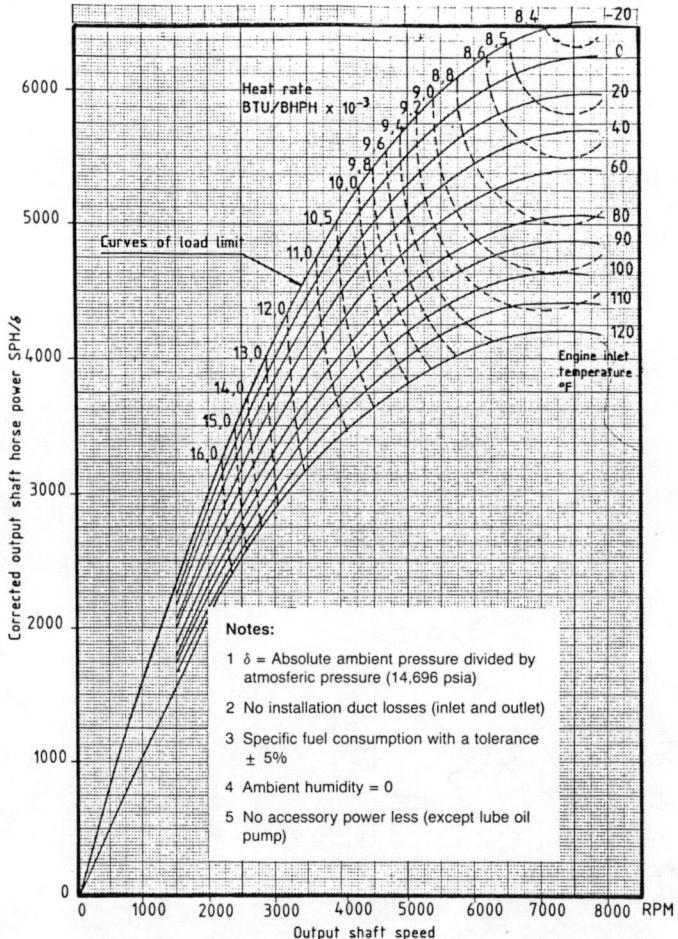

Figure 4-46. Characteristic curves for Dresser Clark DJ 50 GA gas turbine with a maximum continuous power turbine speed of 7,200 rpm. (Courtesy of Dresser Clark.)

curve is then almost vertical so that the axial air compressor acts like a positive displacement compressor. Constant efficiency curves are also shown on these characteristic curves (Figures 4-44, 4-45, and 4-46). Some manufacturers also give for their two-shaft machines a group of curves of percent flow at suction temperature vs. the 100% flow.

The power limit curves can be transposed to an overall specification plot that shows a machine's life expectancy, and surge limit vs. ambient temperature and brake power, as do lines A, B, and C, respectively, in Figure 4-47. The life-expectancy lines A relate to the hot parts of the turbine for a given temperature at the outlet of the combustion chamber. Limiting line B reflects the hot parts' resistance to mechanical fatigue at the same life time as the A lines. And surge limit line C corresponds to the speed at which the air compressor enters into surge, and is related to the nominal surge point at 15°C as:

$$\text{corrected surge speed} = \frac{\text{speed of entering surge at 15°C}}{\sqrt{\dfrac{\text{operating suction temperature K}}{288 \text{ K}}}}$$

The point A corresponds to ISO conditions and to a life of 25,000 hours. The thermal efficiency decreases when the power decreases (Figure 4-48), whether at constant speed, as for turbines driving an alternator, or at variable speed, as for turbines driving centrifugal compressors, when

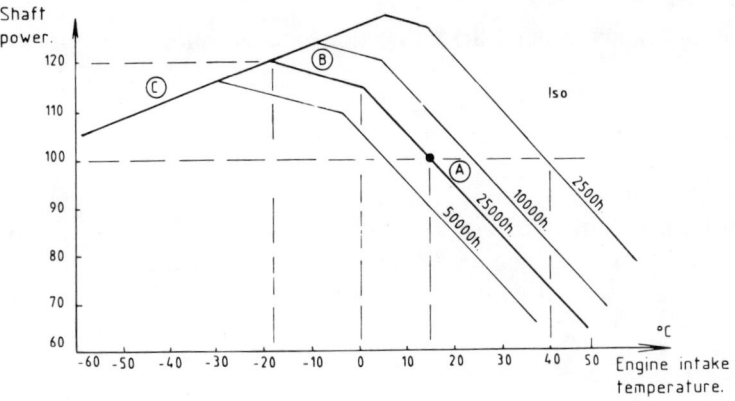

Figure 4-47. Overall specification plot for: (A and B) life expectancy; (C) surge limit of a gas turbine.

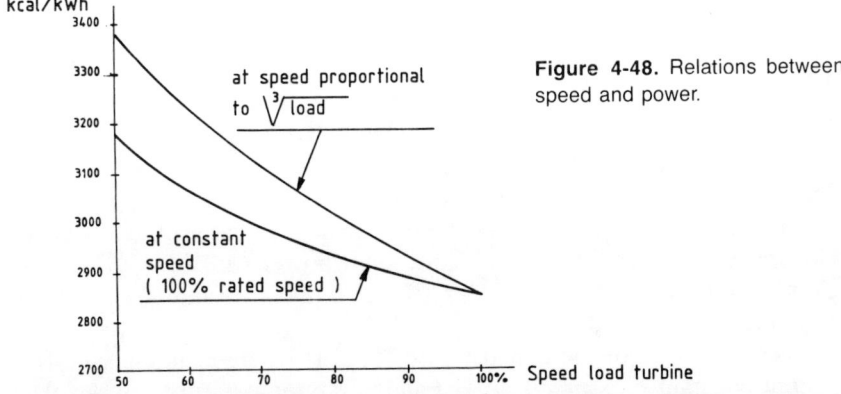

Figure 4-48. Relations between speed and power.

the speed varies as (power)$^{1/3}$. The curves in Figure 4-48 correspond to a second-generation turbine. Figure 3-51 shows that a second-generation gas turbine performs as well as a two-cycle gas engine in constant-torque service.

Exhaust Temperature

Although the exhaust temperature should be indicated by the manufacturer, there will be times when it is desired to calculate that temperature independently. This can be done in two different ways:

- By heat balance around the expansion turbine if the temperature and the pressure of the combustion gases at the outlet of the combustion chamber, the pressure at the outlet of the turbine, and its efficiency are known.
- By heat balance around the entire machine as follows.

Because neither matter nor energy are created nor destroyed within a gas turbine, it is possible to draw an imaginary envelope around the machine and account for all energy that enters or leaves the machine. Considering a gas turbine with a power of W kW, an equivalent dry air weight flow of I_d kg/sec, a specific fuel consumption of Cs kcal/kWh reported to low heating value (LHV), and an air intake temperature of t_1 °C, the energy that enters per second with fuel and air is

$$\frac{Cs\ W}{3,600} + I_d \cdot c_{pa} \cdot t_1 \text{ in kcal/sec}$$

where c_{pa} = the specific heat of air at constant pressure (c_{pa} = 0.24 kcal/kg)

The energy which leaves per second is

- The shaft power: $\dfrac{W \cdot 860}{3,600}$ kcal/sec
- The mechanical losses: x% of shaft power (1% plus 2% per each incorporated gear train if any).
- The radiation and convection heat: y% of shaft power (0.3%).
- Eventually the cooling of variable guiding vanes in cases of heavy duty General Electric gas turbines: z% of shaft power (0.8%).
- Combustion losses % of fuel: ℓ = 2% for liquid fuel and 0% for gaseous fuel.
- The sensible heat of the exhaust gases $I_d\ c_p\ t_4$ where c_p is the specific heat at constant pressure (kcal/kg) and t_4 the temperature (°C) of the

exhaust gas. The weight of exhaust gas can be considered equal to the weight of equivalent dry air at the intake.

Thus

$$\frac{C_s \cdot W}{3,600} + 0.24 \, I_d t_1 = \frac{W \cdot 860}{3,600}\left(1 + \frac{x+y+z}{100}\right) + \frac{\ell}{100}\frac{C_s W}{3,600} + I_d \, c_p \, t_4$$

and

$$t_4 = \frac{\dfrac{C_s \cdot W}{3,600}(1-\ell/100) - \dfrac{W \cdot 860}{3,600}\cdot\left(1+\dfrac{x+y+z}{100}\right)}{I_d \, c_p} + \frac{0.24}{c_p} t_1$$

The specific heat at constant pressure of the exhaust gas depends on heat supplied by the fuel

$$Q = \frac{i_3 - i_2}{100 - x} \cdot 100$$

where i_3 = enthalpy of air at the outlet of the combustion chamber
i_2 = enthalpy at the outlet of the air compressor
x = percent of LHV of the heat to be supplied to heat the fuel at the temperature of the combustion chamber outlet

$c_p = 0.270$ for $Q = 166$ kcal/kg air

$c_p = 0.268$ for $Q = 133$

$c_p = 0.266$ for $Q = 111$

If there is water or steam fed to the combustion chamber, it must be included in the energy balance like the fuel or the air.

Operating Flexibility

Figures 4-44 and 4-45 show that turbines with two shafts are more flexible than those with one shaft with respect to operating speed, with two-shaft turbines capable of 60%–105% of nominal speed compared to 70%–100% for single-shaft turbines. Figure 4-47 shows that operating at more than the 100% nominal power considerably shortens the life of a gas turbine, particularly because that shortens the life of the hot parts of the expansion turbine. This in turn shows that the power of a gas turbine is limited by the maximum temperature of the gases leaving its combustion chamber.

Oil Consumption

Oil consumption is very small. The gas generator for a two-shaft turbine will consume 0.7–1.4 liters per 10,000 kWh; and gas generators derived from aeronautical turboreactors consume only 0.5–1.0 liters per 10,000 kWh. The power turbine and the compressor are lubricated and cooled with the same oil, for which the consumption is on the order of 0.6–0.8 liters per 10,000 kWh.

Startup

A gas turbine cannot operate independently until its speed has achieved 30%–40% of its nominal value. Gas turbines must have some sort of auxiliary power for 1–2 minutes during startup. This power varies from 0.5% of the nominal for a turbine with two shafts to 3% of the nominal for a turbine with one shaft. The power can come from an electric motor, an expansion turbine using fuel gas under pressure or compressed air, an internal combustion reciprocating engine, or even a small gas turbine that is itself started up by an electric motor on batteries, as with aviation gas turbines.

CORRECTIONS FOR OTHER THAN STANDARD OPERATING CONDITIONS

Whether its characteristics refer to ISO conditions, the operation of a gas turbine changes when actual ambient conditions are different from those standards, and it is necessary to calculate predicted operating characteristics for specific turbines in specific locations. This means calculating corrected power, as it is affected by ambient temperature and pressure as well as inlet and outlet pressure drops and ambient humidity, and calculating corrected efficiency as it is affected by the fuel, as well as ambient temperature, pressure, and humidity, and the inlet and outlet presure drops.

Corrected Power

Effects of Atmospheric Pressure and Temperature

Power is proportional to the mass flow, I; mass flow is directly proportional to the absolute pressure, p, and inversely proportional to the absolute temperature, T.

$$I = K_1 \frac{p_1}{T_1} = K_2 \frac{p_3}{\sqrt{T_3}}$$

where p = absolute pressure
T = K

subscripts $_1$ and $_3$ = conditions at the inlet to the air compressor and the expansion turbine, respectively. Power is proportional to the square root of the turbine inlet temperature because flow through the expansion turbine is limited by its Mach number.

For a given temperature at the exit of the combustion chamber, T_3, $(1/T_1)$ is proportional to the expansion ratio (p_3/p_1) of the turbine, and a change in temperature will act proportionally on both the mass flow and the expansion ratio of the turbine, and thus to the same proportion squared on the power, W. Experience has shown that this latter proportion is not exactly the theoretical second power but rather 2.7, as:

$$\frac{W'}{W} = \frac{p_1'}{p_1}$$

$$\frac{W'}{W} = \left(\frac{T_1}{T_1'}\right)^2$$

$$\left(\text{which is in experience } \left(\frac{T_1}{T_1'}\right)^{2.7}\right)$$

or

$$\frac{\Delta W}{W} = -2.7 \frac{\Delta T_1}{T_1}$$

Figure 4-49 shows the relation between ambient pressure and the elevation above sea level.

Discharge Pressure Drop

Because the gases passing through the expansion turbine go from supersonic to subsonic velocities, the Mach number at the inlet determines the flow and separates the pressures at the air compressor's inlet and outlet from effects of varying pressure drop at the turbine's exhaust. The power is proportional to changes in the outlet pressure only if efficiency is 100%, and if Q_3 is the volume flow through the turbine (Q_3 measured at the inlet):

168 Drivers for Rotating Equipment

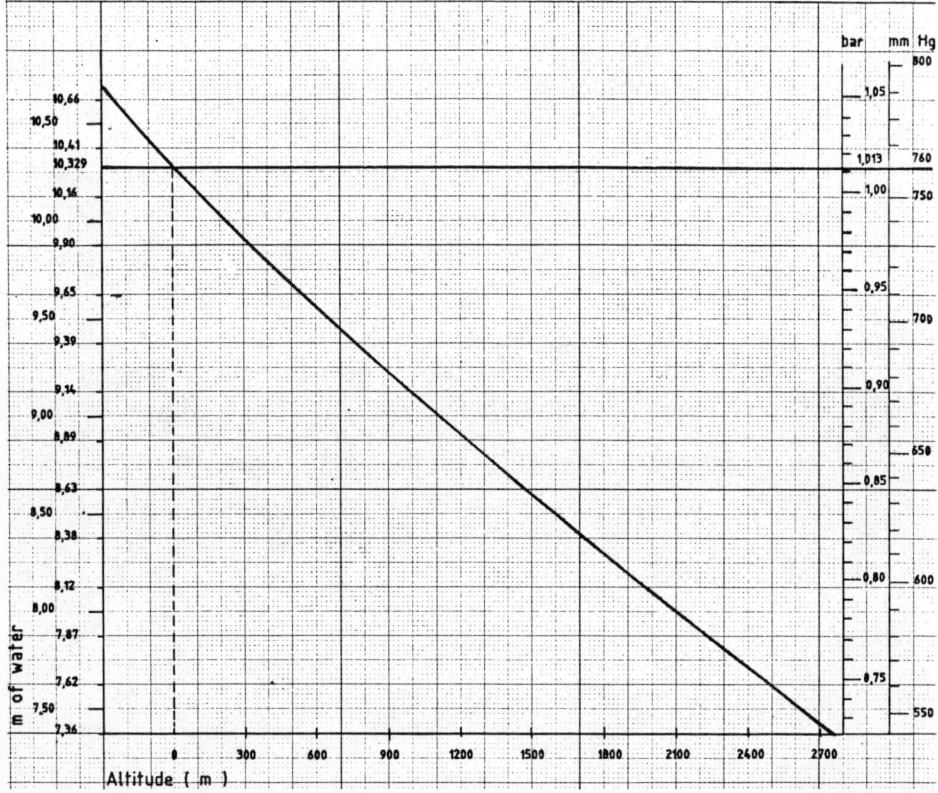

Figure 4-49. Conversions between altitude (m), head of water (m), pressure (bar), and head of mercury (mm Hg).

$$W = p_3 \, Q_3 \, \frac{\gamma}{\gamma - 1} \frac{T_4}{T_3} \left[\left(\frac{p_3}{p_4}\right)^{\frac{\gamma-1}{\gamma}} - 1 \right]$$

$$\frac{\Delta W}{W} = \frac{\Delta \frac{p_3}{p_4}}{\frac{p_3}{p_4}} = -\frac{\Delta p_4}{p_4}$$

Furthermore, an increase in exhaust pressure translates to an increase in the density of the exhaust gases, with a reduction in their outlet velocity and a corresponding increase in efficiency of the expansion turbine, all of which is also reflected in an increase in the exhaust gas temperature because the expansion ratio is lower, although this effect is relatively negligible on the density. In sum, the actual proportion between power and exhaust pressure drop is modified by a factor, k, that is 0.55, 0.62, and 0.75 for suction temperatures of $-30°C$, $15°C$, and $50°C$, respectively.

$$\frac{\Delta W}{W} = -k\frac{\Delta p_4}{p_4} \quad \text{instead of} \quad \frac{\Delta W}{W} = -\frac{\Delta p_4}{p_4} \quad (\Delta p_4 > 0)$$

Suction Pressure Drop

A change in suction pressure at a given exhaust pressure affects all pressure levels proportionally from the inlet to the compressor through the inlet of the expansion turbine. That change thus acts proportionally on the mass flow and the expansion ratio, which affects the power as:

$$\frac{W'}{W} = \left(\frac{p_1'}{p_1}\right)^2$$

However the effects of sonic velocity and possible Mach No., as well as changes in gas density with pressure, combine to change this proportionality as they do that for exhaust pressure. A correction factor, k', is introduced with values of 1.62, 1.72, and 1.82 for inlet temperatures of $-30°C$, $15°C$, and $50°C$, respectively, and:

$$\frac{\Delta W}{W} = k'\frac{\Delta p_1}{p_1} \quad \text{instead of} \quad 2\frac{\Delta p_1}{p_1} \quad (\Delta p_1 < 0)$$

The suction pressure drop acts on the turbine's expansion ratio as does the exhaust pressure drop, and the exhaust temperature is such that:

$$\frac{T_3}{T_4} = \left(\frac{p_3}{p_4}\right)^{\frac{\gamma-1}{\gamma} \cdot \frac{1}{\eta_T}}$$

where $\frac{\gamma-1}{\gamma} \cdot \frac{1}{\eta_T} = 0.36$ when $\eta_T = 0.8$

Where subscripts $_3$ and $_4$ represent the inlet and outlet of the expansion turbine, and η_T is the overall efficiency.

For T_3 constant:

$$\frac{\Delta T_4}{T_4} = 0.36 \frac{\Delta p_1}{p_1} \text{ or } 0.36 \frac{\Delta p_4}{p_4}$$

Actually:

$$\frac{\Delta T_4}{T_4} = 0.60 \frac{\Delta p_1}{p_1} \text{ or } 0.60 \frac{\Delta p_4}{p_4}$$

where $p_1 = p_4$ = atmospheric pressure at the inlet of the filter and the outlet of the exhaust. For its turbines, General Electric takes $k = 1$ and $k' = 2.1$. Accordingly a pressure drop of 100 mm = 4 in. of water at the suction brings on a 2.1% reduction in power at the exhaust, a 1% reduction in power, and at the suction or at the exhaust a 1.7°C increase in exhaust temperature.

Suction and Exhaust Pressure Drops Combined

An empirical formula permits calculating the variation in power as a function of the total pressure drop in the inlet filter and at the exhaust, as follows:

$$\frac{\Delta W}{W} = k \times \frac{\Delta p_1 - \Delta p_4}{p_4}$$

where p_1 = pressure at the inlet to the compressor
p_4 = pressure at the outlet of the expansion turbine

and

$$k = 1 + \frac{e \cdot T_4 \cdot g \cdot kg/sec}{2.94 \cdot 10^5 \, W}$$

where e = variation, % of efficiency, for the expansion turbine relative to the % change in power of the gas turbine; $e = 50$ for expansion turbines with 3–4 stages, including the power turbine, and a large load per stage; and $e = 80$ for expansion turbines with 6–8 stages, including the power turbine, and a low load per stage.
T_4 = exhaust temperature, K
kg/sec = air flow
g = efficiency factor = 0.95–0.98
W = power, kW

Under typical conditions, k varies from 1.7 to 2.2.

Effect of the Ambient Humidity

The constant-pressure specific heat of water vapor is approximately twice that of dry air, so that the water content by weight of humid air can be doubled to obtain the dry-air mass flow equivalent to the humid air. For example, if humid air contains 99% dry air and 1% water vapor, its thermal mass equivalent is $99 + 2 = 101\%$ of the thermal mass of the dry air.

The presence of water vapor in the air increases the available power. For this reason water is sprayed into the air stream between the suction filter and air compressor to increase power in hot weather. This has the double effects of increasing the power through increasing the water vapor and reducing the temperature at the inlet to the compressor. The amount of water that can be evaporated into the air can be determined with a psychometric chart (Figure 4-50). On it the temperatures dry and humid before evaporation indicate the relative humidity and the initial water content. However, the relative humidity at the inlet to the compressor should not go over 80% in order to avoid risk of condensation and consequent corrosion in the compressor.

Corrections for Thermal Efficiency

Effects of the Fuel

The heat energy supplied to the combustion air between the air compressor outlet and the expansion turbine inlet is $i_3 - i_2$ where i_3 is the enthalpy at the inlet of the expansion turbine and i_2 at the outlet of the air compressor. The fuel must supply this heat energy and the sensible residual energy to bring itself at the expansion turbine inlet temperature. This sensible residual heat can be given as a percent of low heating value of the fuel versus the outlet temperature of the combustion chamber (t°C)

$x \% = 0.00106 \, (t - 15)^{1.24552}$ for residual oil

$x \% = 0.00147 \, (t - 15)^{1.21028}$ for light oil

$x \% = 0.00181 \, (t - 15)^{1.20891}$ for natural gas

Thus the combustion efficiency is $(100 - x)\%$ because this sensible residual heat can be considered as a loss.

In case of liquid fuels it is necessary to add 2 to x because of incomplete combustion (2% loss).

172 Drivers for Rotating Equipment

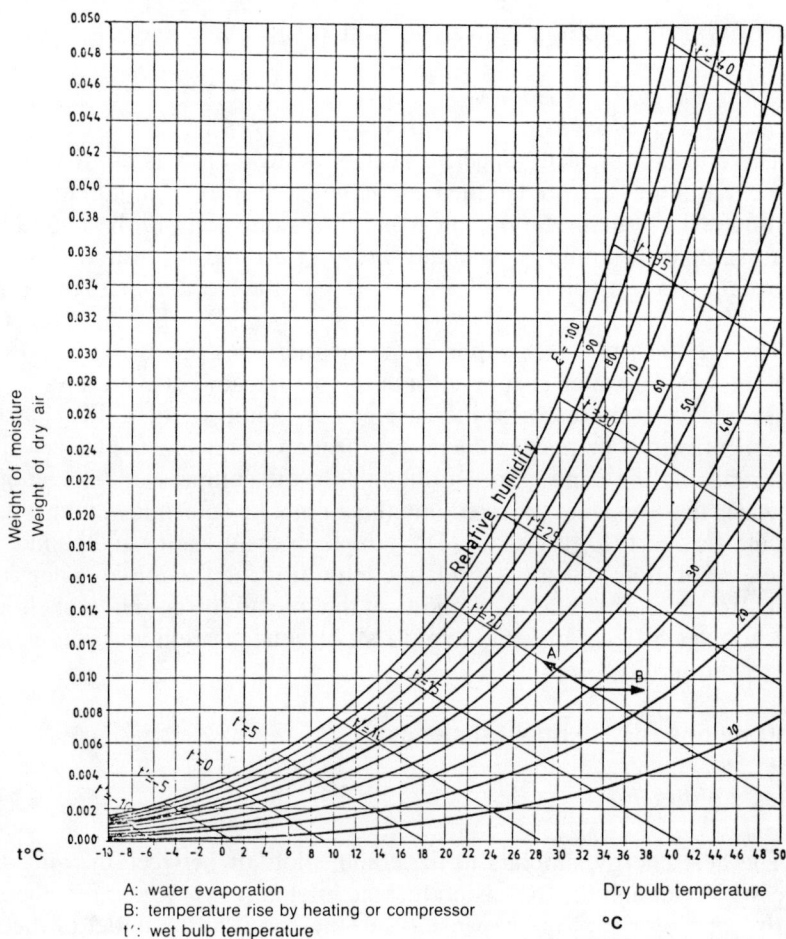

A: water evaporation
B: temperature rise by heating or compressor
t': wet bulb temperature

Dry bulb temperature
°C

Figure 4-50. Psychometric chart.

Thus the heat release rate, to be supplied by the fuel is

$$Q = \frac{i_3 - i_2}{100 - x} \cdot 100 \text{ kcal/kg air}$$

and the thermal efficiency η must be corrected in accordance with the fuel characteristics against $100 - x$ for a reference fuel

$$\frac{\Delta \eta}{\eta} = \frac{\Delta(100 - x)}{100 - x}$$

Gas Turbines

The rate of fuel, kg of fuel per kg of inlet air, is Q/LHV, if LHV is in kcal/kg of fuel and Q in kcal/kg of air. Stoichiometric combustion corresponds to exactly that weight of air required to burn a unit weight of fuel completely to water (H_2O) and carbon dioxide (CO_2); this amounts to 16–17 kg of air per kg of natural gas. Because the air is the working medium of a gas turbine, the turbine will usually take three times as much as the stoichiometric amount of air, or around 50 kg of air per kg of natural gas. The excess air is actual air flow/stoichiometric air flow.

Corrections for Inlet Temperature

For a given compressor-speed and combustion chamber outlet temperature, any change in the gas turbine's inlet temperature tends to change the heat liberated in the combustion chamber, as well as to change the air flow. Both the work of compression and the corresponding enthalpy increase while the temperature (and thus the enthalpy) of the combustion chamber outlet are held constant. However, a change of inlet temperature will be passed along to the chamber outlet, unless the amount of heat liberated in the chamber is changed to compensate. The change in amount of fuel for this is calculated as:

$$\Delta Q = -\frac{\Delta i_1}{100-x} = -\frac{c_p}{100-x} \Delta T_1$$

where ΔT_1 = change in inlet temperature
x = sensible residual heat of fuel to bring it to the combustion chamber outlet temperature
Δi_1 = change in enthalpy corresponding to ΔT_1
c_p = specific heat of air at constant pressure
ΔQ = change in heating value of the fuel

The term $(100 - x)$ may be assumed equal to 95 (93.5 – 96.5) and C_p as equal to 0.24, without significant error. This allows the equation to be reduced to:

$$\Delta Q = -0.255 \, \Delta T$$

In this, ΔQ is in kcal/kg, if ΔT is in °C.

An increase in temperature at the compressor's inlet reduces the air flow by ΔI kg/sec. Because

$$\frac{\Delta I}{I} = -\frac{\Delta T}{T}$$

power is also changed, as:

$$\frac{\Delta W}{W} = -2.7 \frac{\Delta T}{T}$$

as already seen and the reduction in fuel is thus:

$$\left(\frac{\Delta Q}{Q} + \frac{\Delta I}{I}\right) \times 100 \ (\%)$$

And the reduction in power is . . .

$$\frac{\Delta W}{W} \times 100 = -2.7 \frac{\Delta T}{T} \times 100 \ (\%)$$

Thus the increase in inlet temperature has a negative effect on the thermal efficiency because the reduction in power is greater than the corresponding reduction in fuel consumption. When the temperature at the intake of the compressor is lowered, the power and the thermal efficiency are both increased.

Assume for example that the quantity of heat to be furnished, Q, is 140 kcal/kg of air, and that there is an increase of 1°C in temperature relative to ISO conditions of 288 K. The reduction in heat supplied (LHV) is as follows:

$$\left(\frac{0.255}{140} + \frac{1}{288}\right) \times 100 = 0.53\%$$

The reduction in power is 0.94%. And the thermal efficiency is reduced $(0.94 - 0.53) = 0.41\%$ per °C.

Corrections for Pressure

A change in the atmospheric pressure (intake pressure = exhaust pressure) affects only the power by changing the mass flow; it has no effect on the thermal efficiency.

Corrections for Humidity

Because water vapor has a higher thermal mass than air, it changes the power without affecting the thermal efficiency. If v is the water content in 1 + v humid air mass flow, then to calculate the power the humid air mass 1 + v must be replaced by 1 + 2v which is the dry air mass flow equivalent.

Corrections for Inlet and Outlet Pressure Drops

A higher pressure drop at the exhaust reduces the power without affecting the mass flow for air or any temperatures other than the exhaust temperature; therefore it reduces the efficiency as follows:

$$\frac{\Delta \eta}{\eta} = k \frac{\Delta p_4}{p_4}$$

A higher pressure drop at the intake changes both the mass flow (with no effect on efficiency) and the expansion ratio of the expansion turbine analogously to a higher pressure drop at the exhaust, as follows:

$$\frac{\Delta \eta}{\eta} = k \frac{\Delta p_1}{p_1}$$

Letting this relation's factor, k, be equal to 1.0 will show that a pressure drop of 100 mm of water (4 inches of water), either at the inlet or at the discharge, brings on a reduction in efficiency of 1%.

Sample Calculation for Power and Specific Consumption

Reference conditions for a gas turbine are as follows:

W_0: kW
p_0: 1.013 bar
T_0: 288 K (15°C)
I_0: kg/sec
C_{so}: kcal/kWh specific consumption (LHV)

These occur with dry air and no pressure drop at the inlet and the outlet.
Actual conditions at the site are as follows:

- Dry bulb temperature: t_s
- Wet bulb temperature: t_h
- Suction pressure drop: Δp_1
- Discharge pressure drop: Δp_2
- Altitude, H, to which corresponds atmospheric pressure, p

Calculating Actual Inlet and Outlet Conditions

Figure 4-51 gives the relative humidity and the weight of moisture per kg of dry air. If it is decided to evaporate water into the suction air up to a relative humidity of 80%, the new dry bulb temperature at 80% relative humidity can be determined from Figure 4-50 by tracing the wet bulb temperature, because the wet bulb temperature remains constant. Thus the dry bulb temperature and relative weight of moisture are established through Figure 4-50.

The turbine's suction pressure, p_1, and its discharge pressure, p_4, are determined by subtracting and adding the suction and discharge pressure drops, respectively, from the atmospheric pressure, p.

Then the flow of air to the suction is:

$$I = I_0 \times \frac{p_1}{p_0} \cdot \frac{T_0}{T_1} \text{ kg/s}$$

where $T = K$
$p =$ absolute pressure

and the mass flow equivalent of dry air is calculated from this calculated air flow, I, based on the water content, as follows:

$$I_d = I \left(1 + \frac{v}{v+1}\right)$$

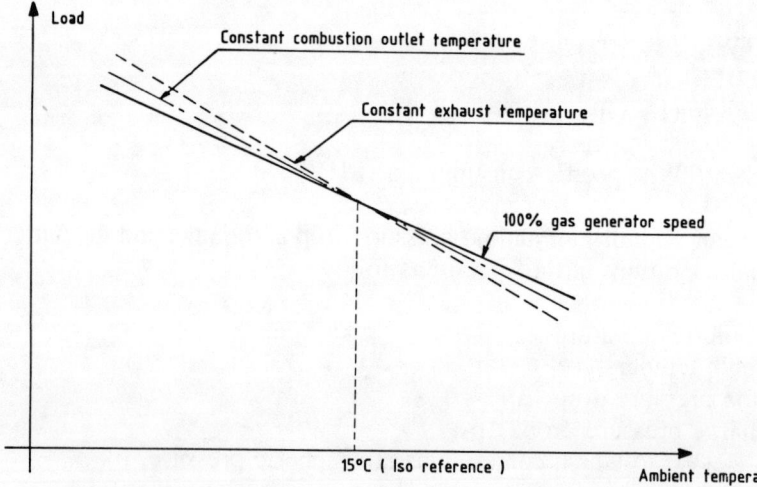

Figure 4-51. Influence of ambient temperature on available power.

$$h = \frac{v}{v+1} = \frac{\text{weight of water}}{\text{weight of wet air}}$$

The weight of water evaporated to have a relative humidity of 80% is calculated as follows:

$$(v_2 - v_1) \cdot I_0 \cdot \frac{p_1}{p_0} \cdot \frac{T_0}{T_1} \cdot (1-h)$$

where v_1 = weight of moisture before water evaporation
v_2 = weight of moisture to 80% relative humidity

The turbine's discharge temperature $T_4 = t_4 + 273$ is calculated as shown earlier.

Calculating the Available Power

The power available at this site is calculated with the following equation, in which the first term represents the temperature correction, the second the pressure correction, the third the correction for pressure drop, and the fourth the correction for humidity.

$$W = W_0 \cdot \left(1 + \frac{T_0 - T_1}{T_0} \cdot 2.7\right) \cdot \frac{p_4}{p_0} \cdot \left(1 - k \frac{|\Delta p_1| + |\Delta p_4|}{p_4}\right)$$
$$\cdot \left(1 + \frac{v}{v+1}\right)$$

In this equation, the factor, k, is calculated as follows:

$$k = 1 + \frac{e \cdot T_4 \cdot I \cdot g}{2.94 \cdot 10^5 \cdot W_0}$$

where e = 50 for expansion turbines with 3–4 stages and 80 for expansion turbines with 6–8 stages. The factor for any eventual step-down gear, g, is equal to 95%–98%.

Fuel Consumption

Fuel consumption at the site is given as follows:

$$\text{kcal/h} = \left[C_{so} \times \frac{W_0}{I_0 \times 3{,}600} + 0.255\,(T_0 - T_1)\right] \times I \times 3{,}600 \times \left(1 + \frac{v}{v+1}\right)$$

And the specific consumption is calculated from this as follows:

$$C_s = \frac{\left[C_{so} \times \dfrac{W_0}{I_0 \times 3{,}600} + 0.255\,(T_0 - T_1)\right] \times I \times 3{,}600 \times \left(1 + \dfrac{v}{v+1}\right)}{W_0 \times \left[1 + 2.7\,(T_0 - T_1)\right] \times \dfrac{p_4}{p_0} \times \left(1 - k\,\dfrac{|\Delta p_1| + |\Delta p_4|}{p_4}\right) \times \left(1 + \dfrac{v}{v+1}\right)}$$

The thermal efficiency, η, can then be calculated from the specific consumption, C_s, as follows:

$$\eta\% = \frac{860}{C_s} \times 100 \quad (C_s \text{ kcal/kWh})$$

In the event the fuel is changed, it is necessary to correct for the sensible residual heat, and the thermal efficiency should be multiplied by:

$$1 + \frac{x_0 - x_1}{100 - x_0}$$

where x_0 = sensible residual heat of the old fuel, percent of LHV
x_1 = sensible residual heat of the new fuel

The specific consumption should accordingly be divided by the latter term.

Simplified Performance Calculation

Power at the Site

In order to correct the rated performance for local conditions, the rated power should be changed as follows:

- Add the wt. % water in the air to the inlet.
- Multiply by the ratio of the pressure at the entrance to the filter over the rating pressure at the inlet.
- Reduce by 2.7% for each percentage point increase in temperature at the inlet, or by 0.9% for each °C increase at the inlet.
- Reduce by 2% for each 100 mm of water pressure drop at the inlet.
- Reduce by 1% for each 100 mm of water pressure drop at the exhaust.

Specific Consumption

In order to correct a rated performance for local conditions, the rated specific consumption should be changed as follows:

- Increase 1% for each 100 mm of water pressure drop at the inlet.
- Increase 1% for each 100 mm of water pressure drop at the outlet.

Exhaust Temperature

The exhaust temperature should be increased 1.7°C for each 100 mm of water pressure drop at the suction or discharge.

Without the effects of inlet and outlet pressure drop, the exhaust temperature becomes equal to the rated iso conditions corrected by the difference between the ambient site temperature and 15°C.

Sample Calculation

Manufacturers of gas turbines generally provide charts for calculating the power and specific consumptions of their machines at conditions different from rated conditions. However it is good to be able to determine these independently, as follows:

A gas turbine with a five-stage expansion turbine and a 98% efficient step-down gear-driver has the following ratings:

- Power: 4,025 kW ISO (15°C, 1.013 bar)
- Air flow: 19.9 kg/sec (ISO)
- Specific consumption (ISO) 2,915 kcal/kWh
- Altitude at the site: 610 m
- Ambient temperature: 27°C (300 K)
- Pressure drop in the suction filter: 100 mm water
- Pressure drop at the exhaust: 150 mm water

What will be the exhaust temperature, the air flow, the ratio of actual to rated power, the local specific power, the energy consumption, the specific consumption and the thermal efficiency?

Exhaust Temperature at ISO Conditions

- The brake power, kcal/kg of air is:

$$\frac{4{,}025 \times 860}{19.9 \times 3{,}600} \qquad 48.3$$

180 Drivers for Rotating Equipment

- The gear loss per kg of air is:

  $(0.02)(48.3)$ .966

- Mechanical losses are constant, and = 0.5   .5
- Heat losses to radiation and convection: .165

       49.93

- Fuel consumption (LHV):

  $$2,915 \, \frac{4,025}{19.9 \times 3,600}$$ 163.72

- The sensible heat carried out in
  the exhaust gases: 113.84

  The temperature of the exhaust gases

  $(113.84/0.27) + 15 = 437\,°C$ 710 K

- Air flow at site conditions:

  $$I = 19.9 \times \frac{0.932}{1.013} \times \frac{288}{300}$$ 17.58 kg/sec

  K coefficient for total pressure drop:

  $$K = 1 + \frac{70 \times 725 \times 17.58 \times 98}{2.94 \times 10^5 \times 4,025} = 1.074$$

- Local specific pressures: (100 mm water = 0.0098 bar ≈ 0.01 bar):

  $p = 0.942$ atmospheric

  $p_4 = 0.957$ outlet of expansion turbine

  $p_1 = 0.932$ inlet of air compressor

  $$W_{site} = 4.025 \times \frac{100 + \frac{-12}{288} \times 2.7 \times 100}{100} \times \frac{0.957}{1.013}$$

$$\times \frac{100 + 1.074 \frac{(-0.025)}{0.957} \times 100}{100}$$

$W_{site} = 3,280$ kW

- Thermal consumption:

$$\left[2,915 \times \frac{4,025}{19.9 \times 3,600} - 0.255 \times 12 \right] \times 19.9 \times 3,600 \times \frac{0.932}{1.013} \times \frac{288}{300}$$

$= 10,169,298$ kcal/h

- Specific consumption: 3,100 kcal/kWh
- Thermal efficiency: 27.7%

OPERATION AND CONTROL—SUCTION AND EXHAUST SILENCERS AND AIR FILTERS

Startup

The operation of a gas turbine becomes independent only after its gas generator has reached 30%–40% of its nominal speed. Accordingly the startup of gas turbines is done under automatic control, with an auxiliary engine of 0.5%–3.0% of the turbine's nominal power turning over the turbine for the first 1–2 minutes (0.5 for aeroderivative and 3 for heavy-duty one-shaft gas turbines).

Control

The control system, whether pneumatic, hydraulic, electronic, or electric, acts to feed the right amount of fuel, as a function of the power demand at the control point, the discharge pressure, the rotating speed, and the pipeline compressor's acceleration during startup. This control system is complemented by mechanisms to protect the machine by avoiding overspeeds, surging in the air compressor, and overheating at the combustion chamber or elsewhere. This control system restricts the power output of the turbine by limiting either the temperature at the combustion chamber's outlet, the temperature at the power turbine's outlet, or the speed of the gas generator. The restriction of the combustion chamber's outlet temperature represents an independent control function, whereas the restriction of the turbine's exhaust temperature or the gas generator's speed are secondary functions related to the rated conditions and the ambient temperature as well as the combustion chamber's outlet temperature. Control of the gas generator's speed is the simplest. Figure 4-51 shows the relations between power, speeds, and the ambient temperature.

Fuel Gas Supply

The gaseous fuel must be free of liquid droplets that can impinge on the vanes, form deposits there, and burn so as to cause local melting. Consequently some sort of mist separator or coalescer is placed between the gas turbine and a fuel gas pressure reducer. Once expanded, however, the fuel gas does not normally need reheating. All turbines accept very cold gases, and only controllers and some valves might be affected. The fuel gas must be at a pressure higher than the combustion chamber's pressure—therefore higher than the compression ratio of the expansion turbine.

Air Filters and Silencers

Filters and silencers have been described in Chapter 2. Air filters are particularly important to second-generation gas turbines, whose expansion turbines have their vanes cooled with a protective laminar layer of cool air; and this air film can be destroyed by silica dust that will melt and clog the holes forming the air film.

Anti-Icing Systems

Temperature drop due to the expansion of air can lead to the occurrence of frost at two points in gas turbines—at the suction filter and at the inlet to the gas generator.

The air filter is subject to frost when the ambient temperature is near freezing and the air is humid. The expansion across the air filter chills the air and leads to frost that can clog the filter. Also, when drops of moisture occur in the air, they can lead to ice formation. Such frosting tends to occur as the speed of the air through the filter becomes too high; so it is important to choose adequately-sized filters. Positive avoidance of frosting can be had merely by raising the temperature of the inlet air 3°–4°C at the most. This can be done either by recycling part of the exhaust gases or preheating the air in heat exchange.

If exhaust gases are recycled, a slip-stream is recycled from the entrance of the exhaust stack to be discharged at the front of the air filter. However, this technique runs the risk of discharging oil vapor into the inlet air, if a leak occurs in the gas turbine.

If the air is preheated, this is done with heat exchange between the exhaust gases and the air in an exchanger whose design lets it additionally

Figure 4-52. Aims-Donaldson anti-icing moisture separator system with exhaust gas circulating in a heat exchanger moisture separator placed at the front of the filter. (Courtesy of Donaldson Cy, Inc.)

Figure 4-53. Donaldson self-cleaning huff-and-puff filter on a Solar Mars gas turbine driving a direct inlet Solar gas compressor. (Courtesy of Donaldson Cy, Inc.)

function as a coalescer (Figure 4-52). By using a self-cleaning filter, clogging of inlet air filters by ice can be avoided (Figure 4-53).

Even when the air filter does not frost, it is possible for stalactites of ice to form in the plenum chamber at the inlet to the gas generator. Such ice stalactites are particularly dangerous, because they can break off and fly into the air compressor to cause partial destruction. Also, ice can form on the first fixed vane that is used in some gas generators. Generally this ice is controlled by a temperature sensor that detects the ice formation and recycles a purge of hot air from the discharge of the air compressor.

Chapter 5
Free-Piston
Gas Generators

Consider, for purposes of analogy, a diesel or dual-fuel engine with a supercharger. Assume that the engine's exhaust pressure is increased. The engine's power will be reduced, but the turbocharger's power will be increased. If this pattern is carried to the extreme, the engine's power becomes effectively zero, as the engine serves merely to supply heated, high-pressure gas to the turbocharger, which takes on so much power that it becomes a driver to an outside load. The result is another type of gas generator and turbine, such as has been described for gas turbines (Figure 5-1).

Since the diesel or dual-fuel engine now serving as a gas generator does not need a drive shaft to deliver power, it can be designed with "free" pistons and without crankshafts, connecting rods, or rotating shafts (Figure 5-2).

OPERATION OF A FREE-PISTON GAS GENERATOR

Free-piston gas generators are diesel or dual-fuel two-cycle engines (Figure 5-2). As the twin pistons move toward each other and the center of the piston chamber, they cover the exhaust ports and trap the air under compression (2). Fuel enters through an injector (7) and fires in the engine cylinder, driving apart the twin pistons, which in turn drive the compressor cylinders to draw fresh air into the compression cylinders (4) through intake valves (5), until the engine pistons uncover the exhaust ports. At that point, the air spring (3) in the compression cylinders starts

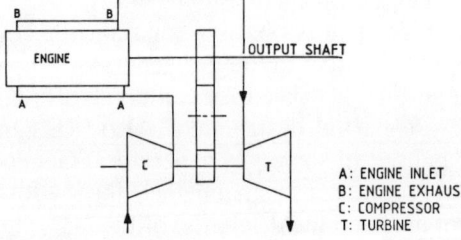

Figure 5-1. Operating principle of a free-piston gas generator with turbine.

A: ENGINE INLET
B: ENGINE EXHAUST
C: COMPRESSOR
T: TURBINE

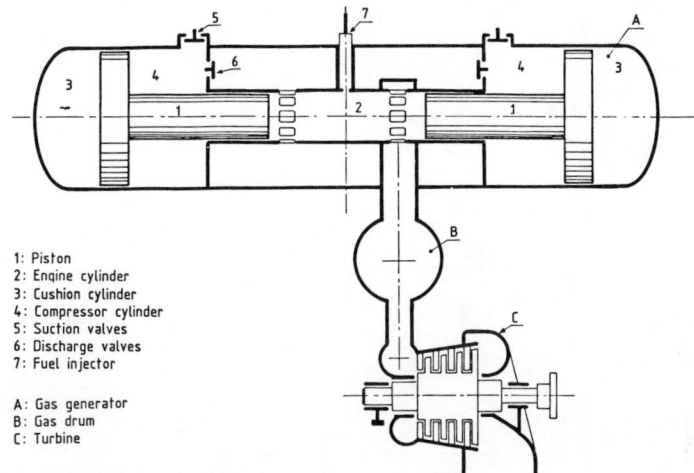

Figure 5-2. Free-piston gas generator feeding a turbine.

1: Piston
2: Engine cylinder
3: Cushion cylinder
4: Compressor cylinder
5: Suction valves
6: Discharge valves
7: Fuel injector

A: Gas generator
B: Gas drum
C: Turbine

the compression piston on its return stroke; and discharge valves (6) open to let out compressed air to chamber B. From there, the compressed air and exhaust gas is expanded through turbine C (Figure 5-2) to recover mechanical energy from the energy of burning fuel injected through valve 7.

OPERATING CHARACTERISTICS

The load-bearing gases leaving a free-piston gas generator have temperatures ranging 450°–500°C at 3.5–4.0 bars pressure gauge (not absolute pressure), depending on operating conditions (Figure 5-3). The maximum pressure in the engine cylinder is 110 bars; and the air temperature in the air-spring remains less than 210°C. Efficiency varies between 20% and 35%, as a function of the operating conditions (Figure 5-3).

Ambient conditions generally affect the available power of a free piston gas generator as a 3% reduction per 500 meters of elevation and a 1% reduction for each 5°C increase in temperature, relative to nominal power.

Because there is no crankshaft, it is difficult to synchronize a spark ignition with engine compression, as well as to obtain a precise air/fuel ratio. Consequently the free-piston engine depends for its ignition on the injection of detonatable gas oil in amounts equivalent to 0.5%–1.0% of the heat release.

186 Drivers for Rotating Equipment

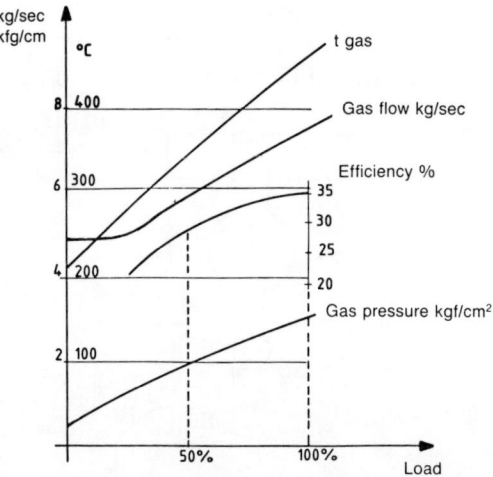

Figure 5-3. A free-piston gas generator. (Courtesy of SN. MAREP.)

Free piston gas generators weigh 13–16 kg/kW, which corresponds to the weights of fast diesel engines of similar power. And they do not take up much space; a generator of 700–1,100 kW would be about 5 meters long and no higher than an average workman (Figure 5-4). Installing the generators requires only connecting the necessary piping, without complicated alignments; and the foundations are small. Although two or more free-piston generators can theoretically feed one expansion turbine, such parallel operation tends to encounter difficulties due to pulsations in the piping manifolds.

Maintenance of these machines is easier than for a diesel engine, but still nevertheless important. In addition to the routine daily maintenance, the machine should be checked over every 1,000 hours, particularly the injectors and valves; at every 4,000 hours it should be dismantled and worn parts replaced; and at every 8,000 hours, it should be given a careful review for replacing rings, valves, pistons and liners.

If a free-piston gas generator always operates at the same speed, a lack of careful lubrication can lead to seizing because of the build up of a lacquer on the cylinder walls, caused by oxidation of fuel and lubricant. Otherwise the lack of a shaft line eliminates the problems of lining up and balancing these machines. And since the expansion turbines fed by these machines are not very hot, the lives of these systems are practically unlimited.

Figure 5-4. Operating conditions typical of free-piston gas generators.

EXISTING FREE-PISTON GAS GENERATORS

Current machines deliver 700–1,100 kW shaft power from the expansion turbine at an overall efficiency of 35%. And several gas generators can be grouped on one turbine, so that the combined power can be as high as 8,000 kW.

Free-piston gas generators have already found places in electrical power-generating stations, ship propulsion systems, and railroad locomotives. However they have not yet found a place in gas transportation, as is theoretically possible with the power turbine driving a centrifugal gas compressor. This is largely because the free-piston engines can hardly operate on solely natural gas that is the only fuel generally available at compressor stations. Add to that the higher efficiency of a gas engine with supercharging, and the questionable edge with respect to investment and maintenance costs, and the free-piston generator is not yet competitive.

Chapter 6
Steam Turbines—
The Rankine Cycle

Steam-turbine-driven centrifugal compressors exhibit wide operating flexibilities. However, the combination of a steam turbine driving a centrifugal compressor is generally not economical, if it requires generating steam solely for that purpose. A steam power system based on a 50-bar boiler operating at 420°C with 10,000–50,000 kW capacity through one or more turbines of 8,000–12,000 kW operating in parallel, with several vacuum condensers in parallel, and with water heaters in series with the condensers to reheat the boiler feed water to 140°C may be capable of an efficiency of about 30% (LHV), so that this system will rarely be competitive with either gas engines or second-generation gas turbines.

On the other hand, if there is a use for steam heat, so that the steam leaving the power turbines can give up its heat of condensation for a useful purpose instead of to the vacuum condensers' cooling water, it becomes possible to develop highly economical power cycles. Also, the need for mechanical power may occur at a location where the atmosphere is hazardous and neither gas turbines nor gas engines can be tolerated for safety reasons. These situations often occur in plants to liquify natural gas as well as in chemical plants and oil refineries.

The Rankine energy conversion cycle applied to steam turbines is analogous to the gas turbine's cycle, except that the Rankine cycle involves both liquid and gaseous phases, instead of the single gas phase involved in the gas turbine's cycle. The Rankine cycle consists of four steps, as follows:

1. High pressure liquid is boiled and possibly superheated to create steam for expansion.
2. The high pressure steam is expanded through a machine that converts part of its thermal energy into mechanical energy.
3. The expanded steam is condensed back to all liquid.
4. The condensed liquid is pumped up to high pressure again.

In this cycle, the high pressure boiler is analogous to the combustion chamber of the gas turbine, the steam expander to the gas turbine's expansion turbine, the condenser to the ambient atmosphere, and the pump to the gas turbine's air compressor. As with the gas turbine cycle, the steam cycle can be improved in various ways, so as to increase the ratio of mechanical to thermal energy, or the efficiency.

THERMAL EFFICIENCY OF THE RANKINE CYCLE

It is possible to carry out the Rankine steam cycle at pressures so high that water is above its critical point, and the heat of change of phase from water to steam and back again is avoided. Also, the relative amount of heat handled during change of phase can be reduced by withdrawing steam from the expansion turbine at an intermediate pressure and resuperheating that steam with heat that can be mostly recovered as mechanical energy. Such cycles can achieve overall efficiencies of 39% LHV, in which latter case the heating value of the fuel assumes condensation of all the steam produced in the combustion.

Figure 6-1 shows the effect of economizer and superheater steps on the Rankine cycle. The steps of the cycle are shown both on the pressure/volume curve of the steam (left) and on its temperature/entropy curve, as follows: Boiler feed water is adiabatically pressurized to vaporization pressure (1-2). The pressurized liquid is further heated in an economizer (2-3). The liquid is vaporized in the boiler (3-4'). The vaporized steam is superheated (4'-4). And the superheated steam is expanded through the steam turbine (4-5). The increase in area of the cycle diagram produced by lines 4-5 relative to 4'-5 shows the increase in conversion due to superheat. With such superheat, the expansion can be entirely dry, with no droplets of water formed to damage the expansion turbine.

If i represents the enthalpy at any given point in the cycle, the cycle efficiency is given by $(i_4 - i_5)/(i_4 - i_1)$. The thermal efficiency, on the other hand, is the net efficiency delivered by the shaft divided by the

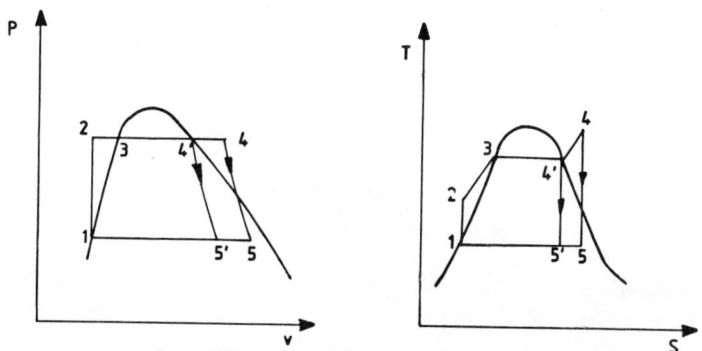

Figure 6-1. The energy conversion cycle for steam turbines.

gross heat input. The net mechanical energy delivered to the shaft takes into account consumption of mechanical energy by feed pumps, boiler recirculating pumps, and fans for the air-cooled condensers. The gross heat input takes account of not only the heat put into the circulating water and steam but also the combustion efficiency of the boiler. This latter efficiency is based on a heat balance around the boiler's firebox and flue-gas heat recovery section, at a data temperature of possibly 15°C (60°F). The heat content, sensible and combustion, of fluids entering is balanced against that of those leaving. Because of their considerable excess air for combustion, gas turbines' exhaust gases still contain much oxygen for combustion and can be used in steam boilers like preheated air, which also enters the boiler's heat balance. Because it is economical to recover over 90% of the heat in steam boilers, their high efficiencies add considerably to combined cycles of gas turbines and steam turbines.

CHOICE OF FLUID

The cycle shown in Figure 6-1 can be applied to an almost infinite number of combinations of fluids, heat sources, and cooling sources; and the widest variety of combinations has evolved in modern chemical plants and petroleum refineries. Plants to manufacture ammonia, for example, compose their large compressor power requirements from a combination of gas turbines, whose exhaust gases feed reforming furnaces, whose hot reaction products give up sensible heat to boil steam, which is used to drive steam turbines before exhausting into reaction tubes whence it passes with the reaction products to be finally condensed against a boiler feed-water preheater. In many such combinations, the temperatures of vaporization and condensation of water under pressure may not be optimally convenient, so a wide variety of other fluids are used. Thus in ethylene plants the reboiler and reflux heat supplied and recovered over a range of temperatures in a line of distillation towers is put into a complex cycle through use of several working fluids such as ethane and propane, which are expanded through turbines driving refrigeration compressors. Figure 6-2 shows typical thermal efficiencies that can be obtained with ammonia (NH_3), butane (C_4H_{10}), and propane (C_3H_8), as well as with water, over a range of temperatures.

These curves show that, if the temperature of an air-cooled condenser is 60°C and the working temperature of the fluid is 350°–400°C, water makes a good choice. Also water has the following advantages:

- It is nontoxic.
- It is nonflammable.
- A steam water circuit does not need to be absolutely leak-free.

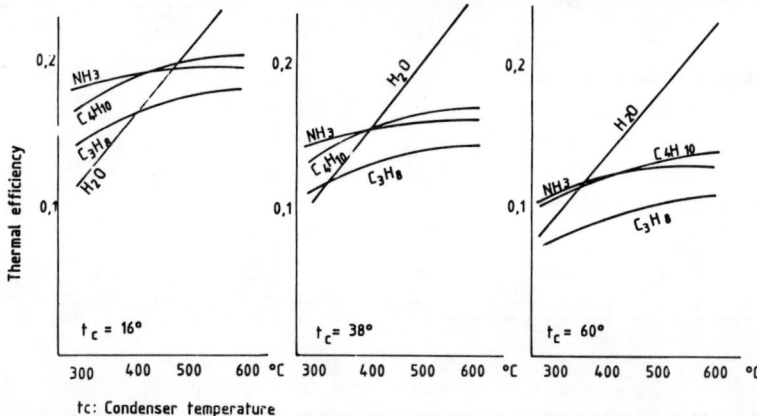

Figure 6-2. Typical thermal efficiencies for turbines operating on different gases.

- Its low cost allows small leaks at the turbine, which in turn allow carbon-ring or simple labyrinth seals.
- Equipment to handle water and steam is widely available and well understood.

On the other hand, a cold climate or a specific operation where the condenser is below 16°C and the temperature of the working gases below 450°C, might make ammonia more attractive. Other considerations are compared as follows:

	H_2O	NH_3	C_3H_8	C_4H_{10}
Safety				
High flammability temperature	+	+	−	−
Limited flammability range with air	+	+	−	−
Less dense than air	+	+	−	−
Ambient vapor pressure higher than ambient pressure	−	+	+	−
Low toxicity	+	−	+	+
Leaks easily detected	−	+	0	0
Operation				
No risk of explosion	+	+	−	−
Low cycle pressure	+	−	−	0

(table continued on next page)

	H_2O	NH_3	C_3H_8	C_4H_{10}
Suitable for heating by gas turbine exhaust	+	+	−	−
Operating conditions				
Noncorrosive	+	+	+	+
No risk of freezing	−	+	+	+
Availability				
Low cost	+	+	+	+
Stable	+	0	0	0

STEAM GENERATION IN THE STEAM-GAS TURBINE COMBINED CYCLE

When the heat in the exhaust from gas turbines is recovered in a boiler without additional firing, the temperature of the hot gases does not exceed 500°C. This limits the available temperature difference, which in turn favors finned tubes and reduced liquid circulation on the order of 2 meters per sec velocity. Two types of steam generators are current, conventional generators with a steam drum, and once-through generators without a steam drum.

Steam Generation at One Pressure

Steam generators operating at a single pressure level can be either the conventional type with a steam drum, or the once-through type without a steam drum.

Steam Generation with Steam Drum

Figure 6-3 shows this system, where the steam drum with two relief valves serves to receive boiling water from the economizer and deliver saturated steam to the superheater, while it provides surge capacity for the water/steam circulation through the vaporizer. Steam from the superheater passes through a safety and a throttle valve directly to the turbine, where the steam expands to give up heat to mechanical energy. The expanded steam then passes to the condenser, which may be water cooled but is generally air cooled and shown as such. Because of the limited hot-gas temperature in the boiler, the steam will not have had sufficient heat to pass through the turbine dry; it will have partially condensed at the turbine exhaust.

Steam Turbines—The Rankine Cycle 193

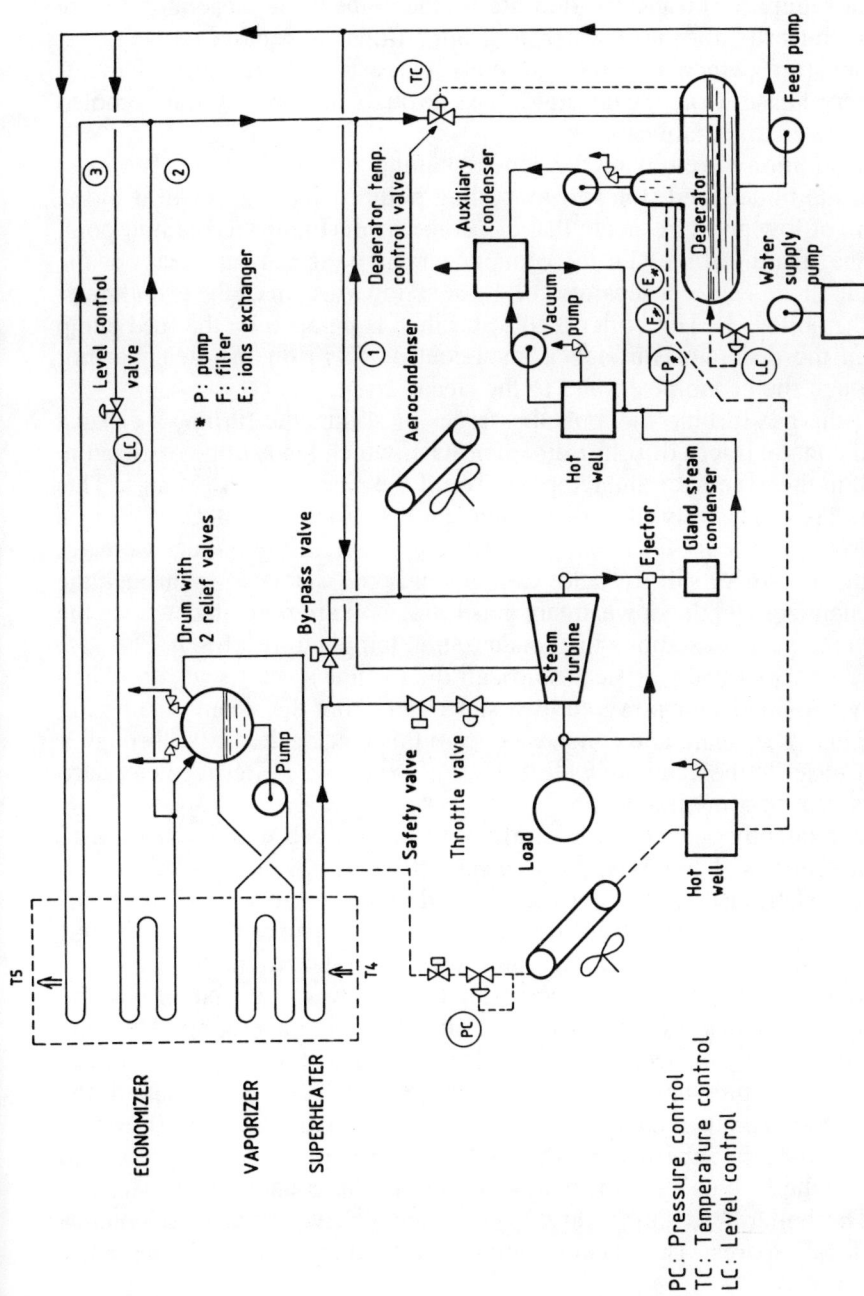

Figure 6-3. Steam power system with a steam drum.

Water from the condenser may run to a hydraulic leg, which is a vertical pipe standing in a "hot well." From the hot well a condensate pump (P in Figure 6-3) transfers the water to the dome of the deaerator, where it, with fresh boiler feed-water, is steam-stripped to remove oxygen-containing air, which can be extremely corrosive in this environment. Stripped gases from the deaerator pass through an auxiliary water-cooled condenser to the atmosphere.

In addition to removing oxygen-containing air, the deaerator serves to heat the boiler feed water to its boiling point. This requires heat in the form of low pressure steam that can come from (Figure 6-3) a mid-point of the steam turbine (1), the economizer via a take-off upstream of the steam drum (2), or a separate coil located downstream of the economizer in the boiler (3). Hot boiler feed water then is pumped by the feed pump from the deaerator through a level-control valve on the steam drum, through the economizer, and to the steam drum.

If the gas turbine fuel contains traces of sulfur, the turbine's exhaust will contain traces of sulfur dioxide (SO_2), which tends to be oxidized to SO_3 in the excess air and temperatures of the gas turbine's exhaust. This SO_3 has a relatively high dew point, on the order of 150°C when it is present in excess of 10 ppm; and if it condenses in the boiler, it creates highly corrosive sulfuric acid. Consequently the water-flows through the economizer and the downstream generating coil are regulated to keep the skin temperatures above the condensation temperature of SO_3.

The normal leaks of steam through the turbine's shaft seals are withdrawn to an ejector powered by a slipstream from the steam feeding the turbine. The steam and condensate from this ejector then pass through a condenser to the main hot well. Losses of expansive boiler feedwater are thus kept to a minimum.

It frequently occurs that several gas turbines, each with its own waste heat boiler, will supply steam to a central steam power turbine. In such cases, each waste heat boiler is connected to the circuit through an isolating valve, and is controlled so that it will not be open to that circuit until its operating conditions are the same as those of the circuit. As a gas turbine is started up to meet an increasing load, its waste heat boiler must be able to start up with the main steam circuit already operating. The startingup steam is then sent to a special condenser that does not need to operate at low pressure and thus needs no pump to send condensate to the deaerator. Another means of regulating this startup steam is shown in Figure 6-4; the superheated startup steam is desuperheated and cooled to match the turbine's outlet, before passing to the condenser.

The boiler feed pumps are electric motor-driven centrifugal pumps with full spares. The vacuum pumps are liquid-ring type with electric motor drives.

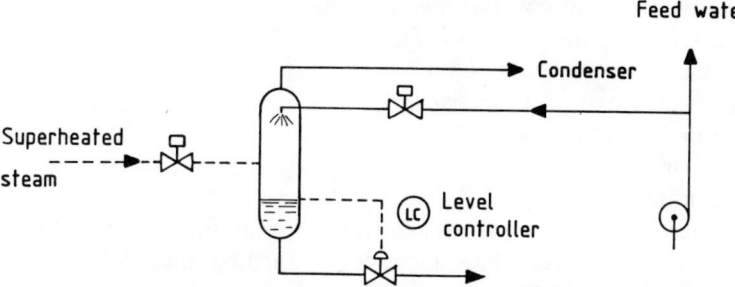

Figure 6-4. Startup steam regulation system.

The quality of boiler feed water must be carefully controlled to keep it from leaving deposits as it is vaporized in the boiler tubes, as well as to prevent corrosion. Deposits are largely prevented through a boiler blowdown, which is calculated as a percentage of the circulation, based on the salt content of the make-up water and the allowable levels in the circulating water. Also, demineralizers and filters are used on the water from the hot well. The corrosiveness of the circulating water is controlled through additions of hydrochloric acid, soda, and hydrazine, with the pH maintained at 6.5–8.5.

The cost of the demineralizers is balanced against the amount of blowdown they can reduce. They are generally filtering beds of ion exchangers operating in parallel, so that one can be isolated for reverse-flow flushing and regeneration while one or more others are on stream.

Control of this steam drum system is through the power steam turbine. If that turbine is coupled to the gas turbine, its throttle valve (Figure 6-3) is wide open and the steam turbine's power varies with that of the gas turbine, as a function of the heat in the gas turbine's exhaust. If the steam turbine is independent of the gas turbine, the steam turbine is controlled through the throttle valve (Figure 6-3). Safety systems include doubled block valves, an automatic turbine shutdown valve, and a bypass from the turbine to the condenser.

Because steam condensate acts as a highly corrosive electrolyte when still, the steam system should be purged with nitrogen if shut down for more than 3–4 weeks.

Typical operating characteristics of waste heat boilers with steam drums are as follows:

- Pressure drop of exhaust gases across heater: 300 mm water
- Exhaust gas temperature:
  In: 420°–500°C
  Out: 220°–240°C

- Steam pressure: 30–50 bar abs.
- Steam temperature: 380°–470°C
- Condensate temperature: 65°C
- Condenser pressure: 0.25 bar abs.
- Deaerator temperature: 105°C
- Deaerator pressure: 1.2 bar abs.
- Pinch point (Figure 6-5): 20°–30°C
- Steam temperature drop between heater outlet and turbine inlet: 5°C
- Pressure drop between heater outlet and turbine inlet: 1 bar
- Hydrochloric acid additions: 30 g/ton of steam
- Soda additions: 30 g/ton of steam
- Hydrazine additions: 0–2 mg/ton of steam

Once-Through Steam Generators

Three characteristic problems are associated with steam drums. They are hazardous, because of the high pressures and temperatures to which they are exposed. They impose a thermal mass and thus a time-response lag on the power system. And they hold up an awkward volume of steam condensate. However, it has been necessary to accept these limitations with fired steam boilers, because the flames in such boilers could overheat tubes and lead to rupture; and the steam protects against such over-

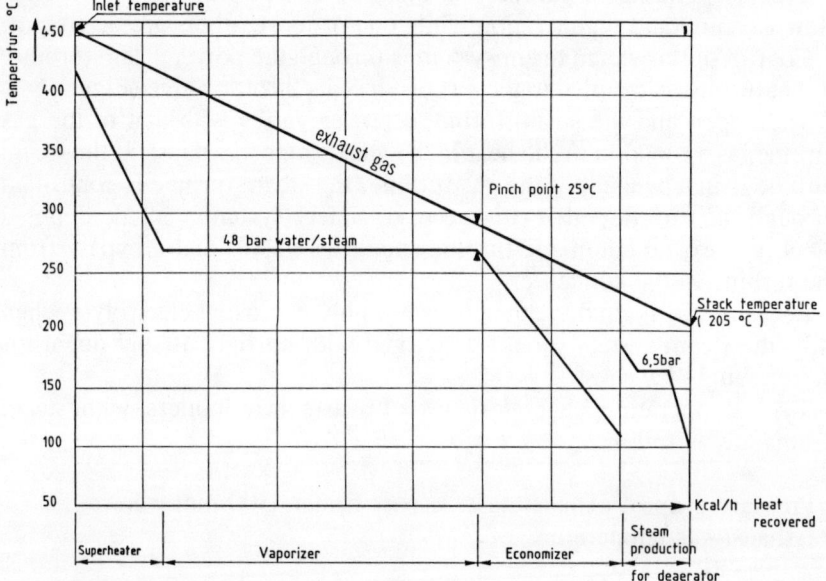

Figure 6-5. Once-through steam power plant with no steam drum.

heating by assuring that there can always be a supply of heat-absorbing water for the tubes.

On the other hand, the exhaust gases from gas turbines do not go above 540°C; and the tubes of waste heat boilers can operate dry below that temperature. It is simply necessary that the tubes be designed to allow for thermal expansion. And on that basis it is possible to design waste heat boilers for once-through operation, with reduced hazard, improved response time, and no hold-up of condensate (Figure 6-6). (Figure 6-6 does not show the deaerator circuit, nor the steam leakage circuit, nor the condenser adapted for several waste heat boilers in parallel.)

These once-through systems use steam generators in which parallel steam tubes are suspended from feed-water manifolds through which the water circulates continuously. The flow through the parallel tubes is balanced by adjustable orifices at their entrances. The point within the tubes at which vaporization is complete and superheating begins is not fixed, but varies with the load. However, the temperature of the steam leaving the boiler controls the supply of boiler feedwater from the deaerator. The temperature measurement is made on a single tube with slightly reduced flow.

Because this system has no steam drum, all water is drained back into the condenser's hot well on shutdown, where it is easily protected against freezing. Also, it is possible to operate this system without full-time supervision, as is legally required for steam drums in some countries.

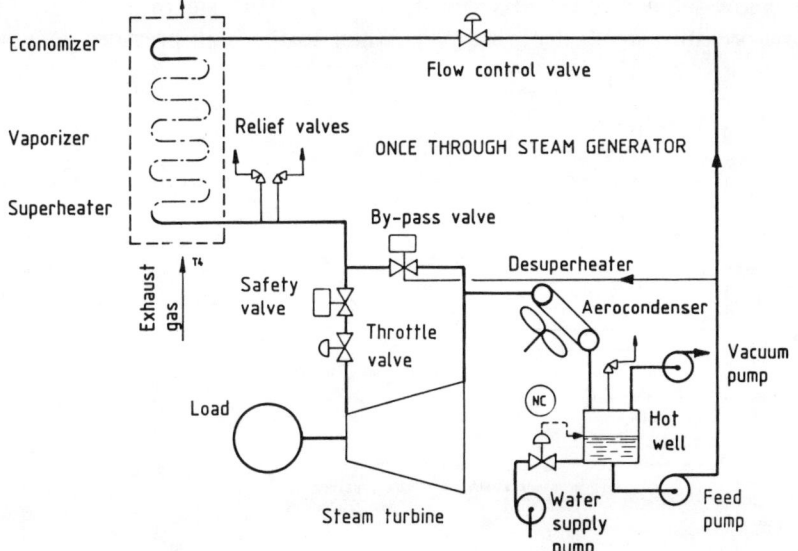

Figure 6-6. Temperature-energy profiles for a steam generator.

Dual-Pressure Steam Systems

Figure 6-5 shows that the maximum amount of steam that can be generated at one pressure is set by a single "pinch point" equal to the minimum economical temperature difference between the hot exhaust gases and the boiling steam. If the pressure of the steam and thus its vaporizing temperature is raised, the heat duty of the vaporizer must be reduced accordingly to find a new 25°C pinch point at a correspondingly higher temperature. This means less steam generation from the same exhaust gases; and less steam generation means a lower thermal-to-mechanical energy conversion efficiency. An important problem thus becomes that of obtaining high pressure from a given gas turbine's exhaust without excessive loss of conversion efficiency.

Figure 6-7 shows the solution to this problem: two pinch points in a dual-pressure steam system. By assigning a smaller duty to the generation of high pressure steam, and another smaller duty to the generation of low pressure steam, the overall duty to steam vaporization is increased, and the maximum steam pressure is raised. The exhaust gas gives up more of its available heat, and the heat recovery efficiency of the waste heat boiler is increased as shown in Figure 6-8.

Both high pressure and low pressure steams are fed to the steam turbine at the appropriate stages (Figure 6-9), so that the turbine's power is increased correspondingly. As shown in Figures 6-7 and 6-9 the low pressure steam is typically not superheated. The economizer at the boiler's low-temperature end serves the low-pressure steam drum. And the low-pressure steam drum supplies water to the high-pressure economi-

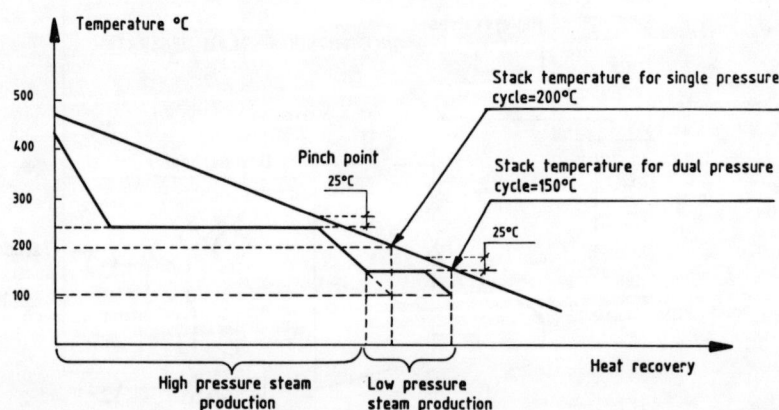

Figure 6-7. Temperature-energy profile for a dual-pressure steam generation with two pinch points.

Steam Turbines—The Rankine Cycle   199

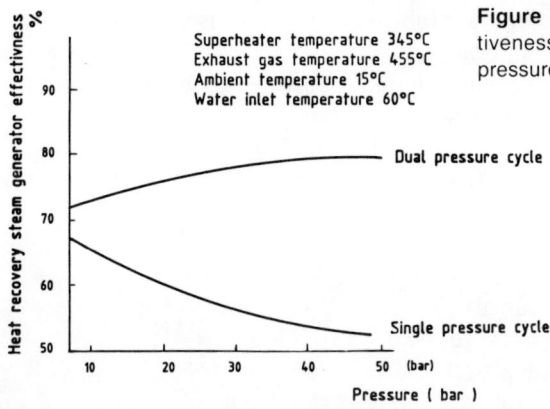

Figure 6-8. Comparison between effectiveness of dual-pressure and single-pressure steam generators.

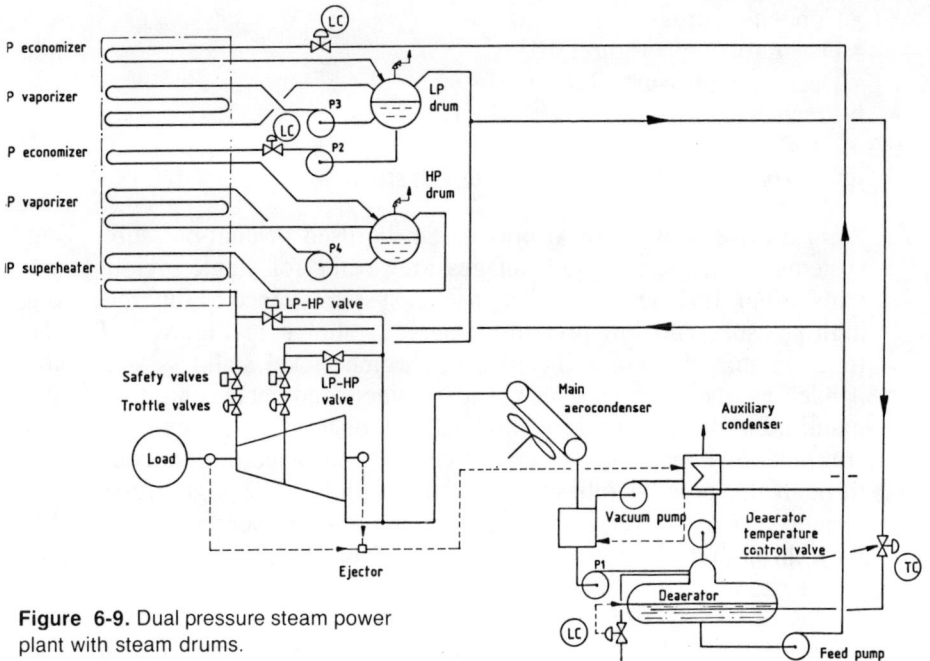

Figure 6-9. Dual pressure steam power plant with steam drums.

zer. Because of the available temperature differences between steam and exhaust gases, it is sometimes possible to use thermosyphoning vaporizers and delete circulating pumps P_4 and P_3 in Figure 6-9. The liquid level in each steam drum is maintained by a level control activating a valve on the economizer feed line.

During startup, the turbine is warmed up with a bypass around the high-pressure shutdown valves. The turbine is brought up to speed with high-pressure steam; and when the high-pressure valve is wide open, the low-pressure valve starts to operate.

Typical operating characteristics of dual-pressure waste heat boilers with two steam drums are as follows:

- Exhaust gas pressure drop: 300 mm H_2O
- Exhaust gas temperatures:
   In: 420°–500°C
   Out: 150°C
- High steam pressure: 35 bar abs.
- High pressure steam temperature: 380°–470°C
- Low steam pressure: 5 bar abs.
- Low pressure steam temperature: 150°C
- Condensate temperature: 60°C
- Condenser pressure: 0.2 bar abs.
- Deaerator temperature: 105°C
- Deaerator pressure: 1.2 bar abs.
- Pinch point difference: 20°–30°C

Once-Through Dual-Pressure Steam Systems

Once-through waste heat boilers can be used in dual-pressure steam systems, with the same advantages they offer for single-pressure systems. Solar Turbines, Inc. offers such a system refined to superheat both high-pressure and low-pressure steam, as shown in Figure 6-10. The tubes in this boiler are finned iron-chrome-nickel stainless steel, suspended by their top ends in vertical layers, and each connected to the manifold at the bottom through flexible coils to allow for expansion and contraction. Each tube has an orifice at its entrance to give equal flow through the parallel tubes out of the manifold. The high-pressure exchange surface is oversized, relative to the low-pressure surface; and the flow through the high-pressure system is four times that of the low-pressure system.

The use of stainless steel tubes in this system eliminates the need for a deaerator, careful pH control, chemical additions for anticorrosion, and periodical purges. By contrast to these stainless steel tubes, mild steel tubes are attacked more by the pure water of steam condensate than by oxygen, because of the electrolytic properties of the condensate. Also, mild steel is attacked by the hot combustion gases, as well as by the sulfuric acid that can form through the condensation of SO_3 at lower temperatures.

The condenser in Figure 6-10 can be air cooled but is shown as water cooled; it has a well with enough hold-up to provide surge capacity for

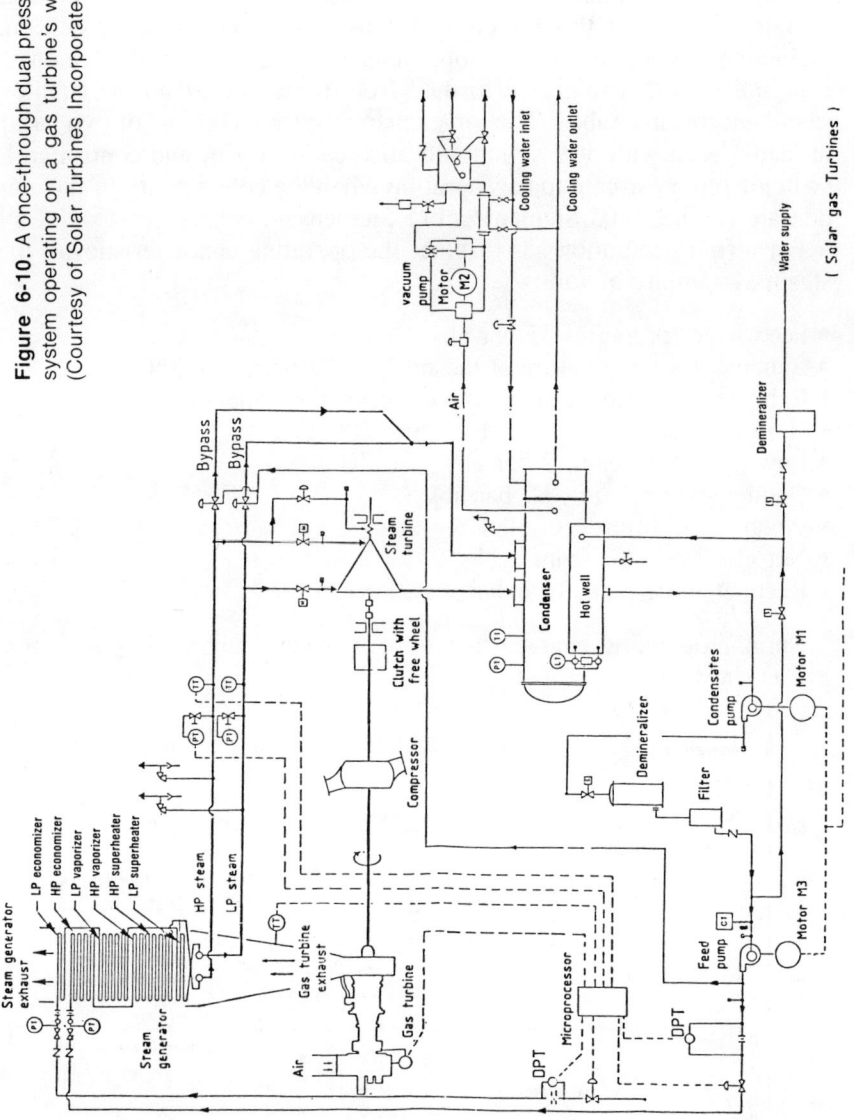

Figure 6-10. A once-through dual pressure steam power system operating on a gas turbine's waste heat boiler. (Courtesy of Solar Turbines Incorporated.)

flow variations and interruptions in the supply of make-up water. The steam turbine's shaft seals are mechanical seals with spring-backed carbon rings; and steam leakage through these seals is drawn off through a high-pressure steam ejector to a water-cooled condenser, from which the condensate is returned to the main condenser's well. The vacuum pump's exhaust from the main condenser well goes to the auxiliary condenser.

Water quality for this boiler is monitored by a conductivity cell and maintained by a filter and an ion exchange system. The filter removes iron and salts that might be formed from parts other than the stainless steel heat transfer tubes. The ion exchange system consists of two parallel beds, each with 50% anion and 50% cation resin, and controlled to switch from on-stream to regeneration when the conductivity of the condensate reaches 0.05 Siemens/cm (1 Siemens=1 mho).

For a first-generation gas turbine, the operating characteristics of this steam system are as follows:

- Feed water pressure: 35 bar abs.
- Exhaust gas temperature at the inlet to the boiler: 445°C
- Exhaust gas temperature at the outlet of the boiler: 60°C
- High pressure steam: 15.5 bar abs., 400°C
- Low pressure steam: 2 bar abs., 215°C
- Condenser pressure: 0.2 bar abs.
- Pinch point difference: 30°C
- Net efficiency of steam cycle: 23%
- Recovery efficiency of exhaust boiler: 80.7%

Similar operating characteristics for a second-generation gas turbine are as shown in Figure 6-11.

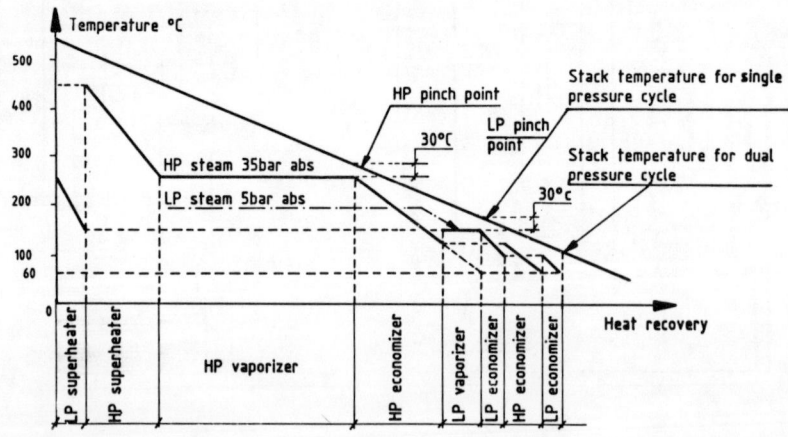

Figure 6-11. Temperature-energy profile for a once-through dual-pressure steam generator operating on the exhaust of a second-generation gas turbine.

Operating Problems of Once-Through Steam Boilers

Because the working fluid enters the tubes of once-through boilers as liquid and appears at the outlet as superheated vapor, the point of vaporization is not fixed. This leads to two problems characteristic of such boilers. First, there is a possibility of producing liquid at the outlets of the tubes during startup; and second, there is a possibility that the flows through the parallel tubes would not be equal. The second problem is particularly important with working fluids such as ammonia that might be dissociated at high temperatures.

The problem of liquid at the boiler outlet can be resolved by locating vapor-liquid separators at the boiler outlets, so as to prevent any liquid from being carried to the steam turbine inlet.

The problem of liquid carryover due to unequal vaporization in the tubes is augmented by the natural tendency of the exhaust gases to flow from bottom to top. Respecting the pinch-points in a most efficient heat transfer system (Figure 6-11) means exchanging low-temperature exhaust for water in the economizers; and if the gases flow from bottom to top, the water-filled tubes are placed above the superheated tubes. The tendency for this water to descend to the superheating tubes requires extra care in sizing the distribution orifices and tube lengths. A means of resolving this problem is to have the exhaust gases flow from top to bottom by means of dampers in the stack (Figure 6-12). Thus the economizers are located at the bottom with the superheating tubes at the top; and if this flow pattern is used with relatively long tubes, the calibrated distributing orifices can be eliminated entirely. One example of a system with both liquid-vapor phase separators and top-to-bottom gas flow is shown in Figure 6-13.

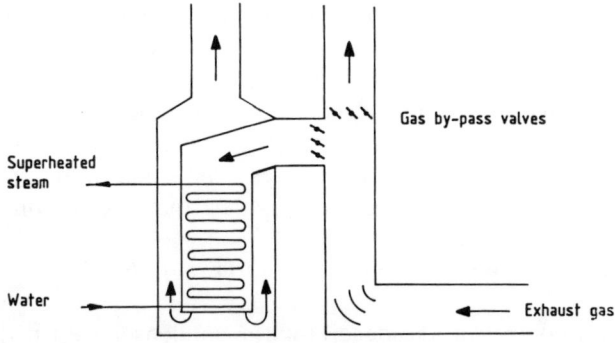

Figure 6-12. Vertical downflow of stack gases to avoid liquid carryover.

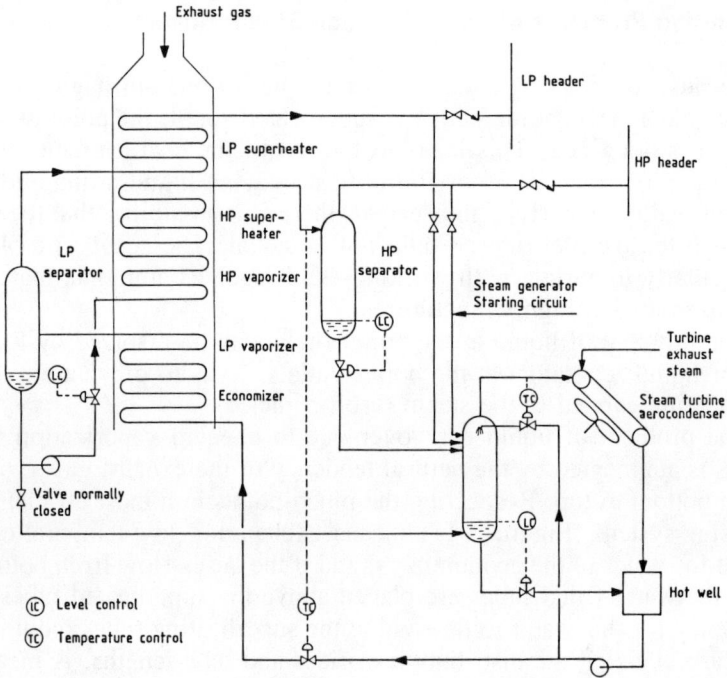

Figure 6-13. Liquid carryover eliminated by vertical stack-gas downflow plus vapor-liquid separators.

Regulation and Control

The regulation of waste-heat power generators is entirely automatic by means of microcomputers. These computers have programs for startup and shutdown of the gas turbine, steam cycle, and driven machine, plus programs based on the gas turbine demand and on the relative demands of the steam turbine if they drive two different machines. When the steam turbine supplies all its power to the same driven machine as the gas turbine, the system control point acts on the gas turbine alone, and the steam turbine operates wide open through a free-wheel coupling. In this case a variation in the gas turbine's load causes a variation in the pressure and temperature of the superheated steam, since there is no control on the inlet to the steam turbine.

The secondary control loop consists of (Figure 6-13):

1. A level control on the deaerator (or condenser well if there is no deaerator) from a treated water makeup tank.
2. A pressure control in the main condenser (through regulation of the cooling air of water). The pressure must be sufficient to avoid a

too-high steam velocity in the steam turbine which must stay under the limit of the sound velocity.
3. A pressure control in the hot well of the startup condenser (if there is one). A startup condenser is necessary if many steam boilers are in parallel.
4. A temperature control in the deaerator on the flow of the reheating steam.
5. A flow control on the feed water either to the steam drums (the set point is the water level in the drums) in a conventional boiler or into the tubes of a once-through steam generator (the set point is the steam temperature at the outlet of the steam generator).

In case of a dual pressure once-through steam generator the flow control

- Maintains constant temperature difference between the high-pressure steam at the outlet of the steam generator and the gas turbine exhaust gas at the inlet of the steam generator.
- Maintains a one-to-four water ratio between high pressure and low pressure circuits.

Depending on the water hold-up, the startup of a gas turbine and centrifugal compressor system requires from 30 minutes for a once-through system to 60 minutes for a system with a steam drum. This startup involves six main steps:

1. Startup of the water circulation, vacuum, lubrication, and compressor seal systems.
2. Startup of the gas turbine.
3. Bringing the centrifugal compressor on stream.
4. Feeding water through the once-through boiler.
5. Heating up the steam turbine.
6. Generating high pressure steam, as follows.

Step 6 is accomplished by starting up the circulating systems, which means starting the pumps, ejectors, vacuum pumps, etc. for the systems, as well as their level controllers, etc. Starting up the gas turbine means letting in the fuel gas, ignition, and bringing the turbine up to its control point (about 90 seconds). Then this turbine is brought on stream to take up its load. After 10 minutes for a once-through system (or 20 minutes for a steam-drum system), the steam pressure is high enough to heat up the steam circuits through their bypass valves; and the steam turbine is heated up through a small bypass on the steam inlet. After about 15 minutes of heating up, the high-pressure inlet valve opens and its bypass closes. Under control of its inlet valve, the steam turbine accelerates rapidly overpassing its first critical speed to the speed at which its clutch on the gas turbine's shaft line causes the steam turbine to take up its load, and the low-pressure steam valves open as those bypass valves close. When the steam turbine begins accepting its full load, both its high-pressure and low-pressure inlet valves are wide open, and its control is made by the heat delivered from the gas turbine's exhaust gases. If the steam turbine's load is separate from the gas turbine's, the steam inlet flows are controlled by set points.

Because the steam cycle's response is slower than the gas turbine's, a reduction in load on the gas turbine leads to a slower reduction in power on the steam turbine, so that it takes 5–10 minutes for the systems to arrive at a new equilibrium.

Shutdown of the combined cycles proceeds through two steps. First, the high-pressure and low-pressure steam inlet valves are closed, as the steam bypass valves and feedwater injection valves are opened to desuperheat the high-pressure and low-pressure systems. Then the gas turbine is shut down. (These two operations are simultaneous in an emergency.) The loads are taken off, but the various circulating pumps as well as the desuperheating water injections are maintained in service for 30 minutes.

A combined cycle that has just been shut down can be started up right away, as long as there is steam pressure and the system is up to temperature. Otherwise, it is necessary to go through the complete startup procedure. However, if the once-through system has operated dry, with the steam turbine unloaded while the gas turbine was operating, it is necessary to wait 30 minutes or so, until the boiler has cooled down, before starting up again.

CALCULATING PERFORMANCE FOR A GAS TURBINE WITH A WASTE HEAT BOILER AND STEAM TURBINE

These calculations involve three separate systems—the gas turbine, its waste heat boiler, and the steam turbine—which are related through heat transfer, thermal-to-mechanical energy conversion, and shared mechanical loads. The calculations are made with the aid of temperature-entropy or Mollier enthalpy-entropy diagrams. Starting with the gas turbine, the calculations proceed as follows.

Gas Turbine Calculations

Mass and energy balances around the gas turbine establish the flows of air, fuel, and exhaust gases, as well as the temperatures, heat contents, and mechanical power delivered. A back pressure of 200 mm water is assumed to allow for the pressure drop of the exhaust gases through the waste heat boiler. The calculations are carried out as described in Chapter 4.

Steam Turbines—The Rankine Cycle

To simplify, the exhaust gas' mass is equal to the inlet air mass (the difference is fuel mass, which is lost by leakages).

Waste Heat Boilers

The heat exchanged in the waste heat boiler is established through a simple heat balance, such as is used on other types of heat exchangers, vaporizers or condensers. Sensible heat carried by the high temperature exhaust gases, on the one side, is absorbed as both sensible heat (in the superheaters and economizers) and heat of change-of-phase (in the vaporizers), as shown in Figure 6-11. The exhaust gas' inlet and outlet temperatures for each heat exchange duty are determined by not only the duty but also the mass flow and the specific heat of the exhaust gases (Figure 6-14), which allows for the calculation of the enthalpies. The calculation of the system is by trial and error and must be started with steam flow estimation (Figure 6-15).

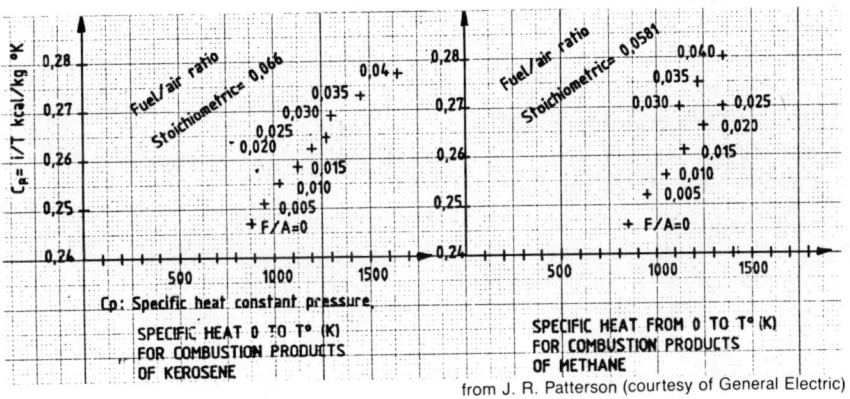

Figure 6-14. Specific heats of exhaust gases.

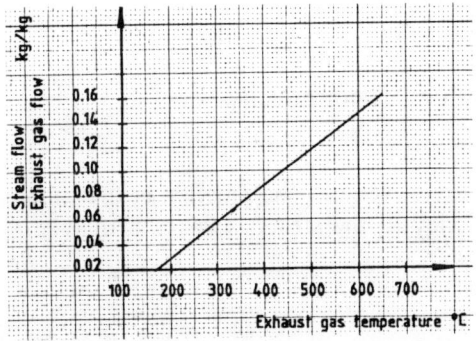

Figure 6-15. Relation of steam flow to exhaust gas flow.

208 Drivers for Rotating Equipment

Sample Calculation

Assume:

1. The pressure drop between superheater outlet and inlet of steam turbine throttle valve is 0.7 bar.
2. The pressure drops of the economizer and superheater are 0.7 bar.
3. The water temperature at the economizer inlet is equal to the temperature in the hot well if there is no deaerator.
4. A condenser pressure of 200 mm Hg (270 mbar) at 30°C (50 mm Hg (70 mbar) at 0°C for an air condenser or 100 mm Hg (150 mbar) for a water condenser).

and assume a gas turbine of the following characteristics:

- Fuel: natural gas, specific gravity 0.6—LHV 8,000 kcal/nm³, 10,335 kcal/kg
- Power: 11,560 kW
- Ambient pressure: 1.013 bar
- Specific consumption: 36,570,000 kcal/h (3,540 kg/h)
- Thermal efficiency (LHV): 27.2%
- Exhaust temperature: 520°C
- Exhaust gas mass flow: 198,500 kg/h
- Fuel/air ratio: 0.02 in mass
- Exhaust gas specific calorific heat at constant pressure: $C_p = 0.27$
- Pressure at the boiler inlet: 30 mbar and a Rankine cycle of following characteristics (Figure 6-16)
- Pressure at the superheater outlet: 41 bar
- Temperature at the superheater outlet: 455°C
- Condenser pressure: 150 mbar (water temperature 55°C)
- Evaporator effectiveness: 0.85

$$0.85 = \frac{T_2 - T_5}{T_2 - T_6} = \frac{\text{temperature difference in exhaust gas}}{\text{temperature difference between exhaust gas inlet and water inlet}}$$

- $T_5 - T_6$ pinch point: function of evaporator effectiveness
- No deaerator

The greater is the evaporator effectiveness, the smaller is the pinch point. The higher the cycle efficiency is, the larger is the exchange surface and the higher the boiler cost.

Steam Turbines—The Rankine Cycle

	Temperature t°C	Pressure bar abs.	Enthalpy Kcal/kg	Flow Kg/h
1	520	1,043	214	198.500
2	? (471)	–	? (201)	198.500
3	455	42	798	? (24.270)
4	255	42,7	670	? (24.270)
5	? (287)	–	? (151)	198.500
6	255	42,7	265,2	? (24.270)
7	? (198)	1,013	? (127)	198.500
8	55	43,4	55	? (24.270)

(Calculated values (from i.S or T.S diagrams and cp diagrams) : _____

(Calculated values from heat balance : ( ))

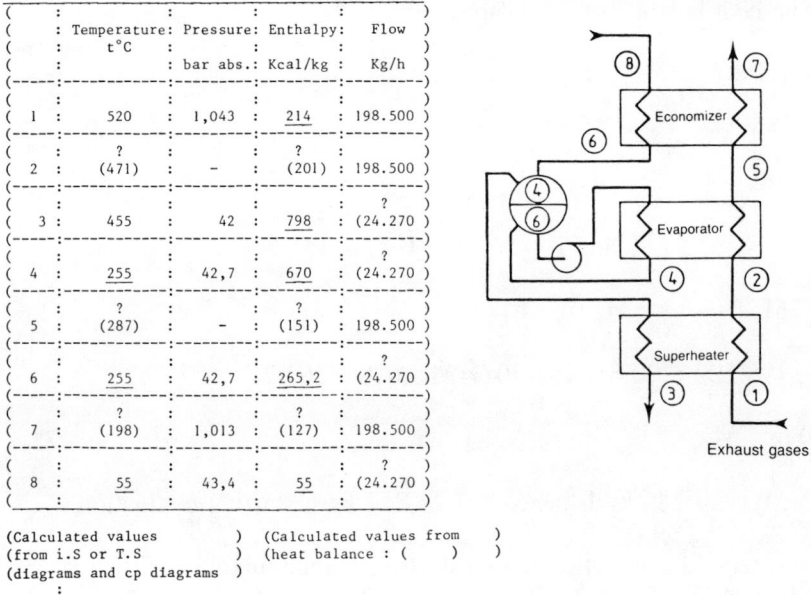

Figure 6-16. Computing heat exchangers in steam generation.

The steam production evaluated to start the calculation is (Figure 6-15):

$$M_2 = 0.122 \times 198{,}500 = 24{,}217 \text{ kg/h}$$

- The heat balance of the superheater is

$$M_1 (i_1 - i_2) = M_2 (i_3 - i_4)$$

$$198{,}500 (214 - i_2) = 24{,}217 (798 - 690)$$

thus

$$i_2 = 214 - 13.18 = 200.82 \quad \left(T_2 = \frac{i_2}{C_p}\right)$$

and

$$T_2 = 743.79 \text{ K } (t_2 = 470.79°C)$$

- The heat balance of the evaporator is

$$\frac{t_2 - t_5}{t_2 - t_6} = 0.85 = \frac{470.8 - t_5}{470.8 - 255}$$

thus

$t_5 = 287.37°C$ and $i_5 = 560 \text{ K} \times 0.27 = 151.30$

$M_1 (i_2 - i_5) = M_2 (i_4 - i_6)$

$198,500 (200.8 - 151.30) = M_2 (670 - 265.2)$

thus

$M_2 = 24,285$ kg/h instead of 24,217 kg/h starting evaluation.

A second calculation with a starting evaluation of 24,000 kg/h gives $M_2 = 24,334$ kg/h; 24,500 gives 24,220 kg/h (Figure 6-17).
The true value is 24,273 kg/h, which gives:

$t_2 = 470.68$ (°C)

$i_2 = 200.79$

$t_5 = 287.35$

$i_5 = 151.30$

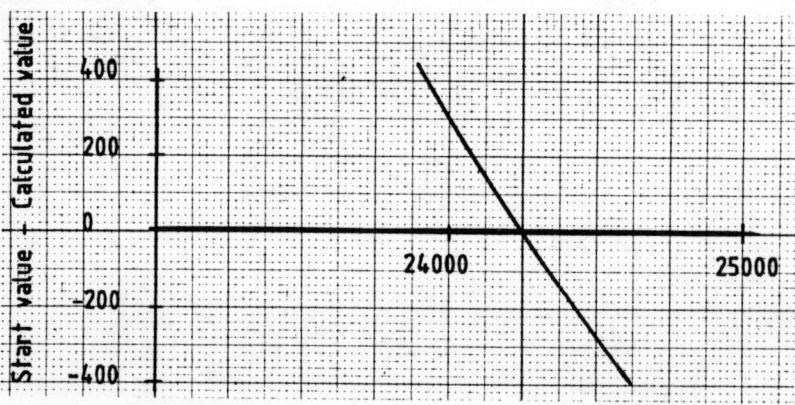

Figure 6-17. Plot for solving trial-and-error calculation of generator duties.

- The heat balance of the economizer is

  $M_1 (i_5 - i_7) = M_2 (i_6 - i_8)$

  $198{,}500 (151.3 - i_7) = 24{,}273 (265.2 - 55)$

  $i_7 = 125.6$

  $T_7 = 465.2 \text{ K} \; (t_7 = 192\,°\text{C})$

- The economizer effectiveness is

  $$\frac{t_5 - t_7}{t_5 - t_8} = \frac{287.35 - 192.2}{287.35 - 55} = 0.41$$

- The superheater effectiveness is

  $$\frac{t_1 - t_2}{t_1 - t_4} = \frac{520 - 471}{520 - 255} = 0.13$$

The evaporator effectiveness has dominance. An easier method is to solve the problem visually by a plot similar to Figure 6-18, where the temperature of each stream is plotted against its heat carried, in kcal/hr. That is, the flow, kg/h, is multiplied by the enthalpy, kcal/kg, to yield kcal/hr. This heat carried is further adjusted to read upward or downward from an arbitrary temperature level, so that comparing plots automatically compares heat duties and temperature differences.

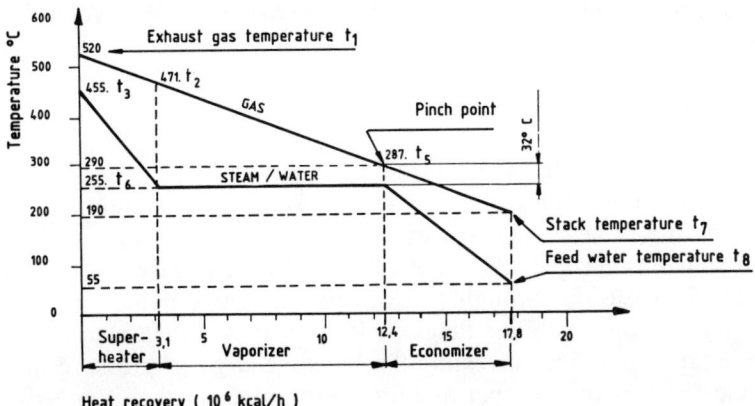

Figure 6-18. Temperature-energy profile for optimizing heat transfer in a steam generator.

Thus in Figure 6-18, the exhaust temperature, $t_1 = 520°C$, is taken as the top temperature level for the exhaust gases; and the heat they carry at any given temperature less than that is plotted versus the amount of heat given up, so that they give: 3.1 mkcal/h by dropping to 471°C, 12.4 mkcal/h by 287°C, and 17.8 mkcal/h by the time they reach the stack at 190°C. This plot thus represents a continuous heat balance on the exhaust gases. Similar plots are constructed for the superheater, vaporizer, and economizer. Because the flows through these components are not fixed, the plot is usually developed through certain relations based on experience. The flow through the superheater, for example, might be taken as that required to keep the steam from crossing its saturation line (on a Mollier diagram) as it expands through the steam turbine; and the flow through the economizer might be taken as the total flow through the steam turbine.

The critical design value in matching these plots is the "pinch point," which is the smallest economical temperature difference between the steam and the exhaust gases. Studies of cost versus efficiency show that costs go up asymptotically, as the pinch point gets below 25°–30°C. And the exact choice of temperature difference depends on local energy costs.

The value t_3 of the superheated steam is determined as that temperature needed to avoid condensation when expanding steam from a pressure corresponding to t_4. These two temperatures, the steam's sensible heat difference between t_4 and t_3 and the latent heat of vaporization, are then determined from a Mollier diagram for steam or from steam tables. Assigning an arbitrary flow of 100 kg/h of vaporizing steam affords a model plot similar to the steam/water lines for the superheater and vaporizer and economizer in Figure 6-18. This arbitrary 100 kg/h is then multiplied to make the model stretch to the pinch point in Figure 6-18, which gives the duties, flows, and temperatures for the system. More complex systems with multiple pinch points (Figure 6-11) can be developed similarly.

Calculating the Steam Turbine

Expansion through a steam turbine is an adiabatic process. That is, the heat added to or taken from outside the turbine is effectively zero; and a zero change in heat, $\Delta Q = 0$, means that the change in entropy is also zero and is isentropic, $\Delta S = (AQ/T) = 0$. Consequently, the theoretical effect of expansion through a steam turbine corresponds to a vertical straight line on a Mollier diagram for steam. However the thermodynamic efficiency for converting the kinetic energy of pressure into mechanical energy of work per unit time, or power, is only 60%–80% for a steam turbine; and this means that the actual line on a Mollier diagram

Steam Turbines—The Rankine Cycle

for "polytropic" expansion through a steam turbine is slanted to give the same pressure but an outlet enthalpy higher than the theoretical by a factor of 1/0.6 to 1/0.7, with a correspondingly higher temperature.

Thus if M_2 is the mass flow through a steam turbine, i_e the entering enthalpy at 40.3 bar and 455°C, i_s the enthalpy of isentropic expansion to 150 mbar and 55°C and 14% condensation, the total duty $Q = 6,286,707$ kcal/h, and the efficiency is 0.75, and $M_2 = 24,273$ kg/h:

- The inlet enthalpy, $i_e = 797$ kcal/kg
- The theoretical outlet enthalpy, $i_s = 538$
- The theoretical power, $M_2 (i_e - i_s) = 6,286.707$ kcal/h $= 7,310$ kW
- The actual power $= (7,310 \times 0.75) = 5,482$.

With the gas turbine and steam generator of the sample calculation, the total shaft power is $11,650 + 5,482 = 17,040$ kW and the gross efficiency of the combined cycle is $17,040 \times 860/36,570,000 = 40\%$.

FEEDING THE WASTE HEAT BOILER

As described earlier, any sulfur in the fuel can become sulfur trioxide (SO_3) in the fuel gases, and this SO_3 can condense on the tubes of the waste heat boiler when the skin temperature is too low. As a rule of thumb, 10% of the sulfur can be assumed converted to SO_3. This rule is used to estimate the SO_3 content of the exhaust gases, and that estimated concentration is used with Figure 6-19 to calculate the corresponding maximum skin temperature, which is in turn calculated from the relative inside and outside heat transfer rates:

$$t_s = t_8 - (t_8 - t_7)(24/240)$$

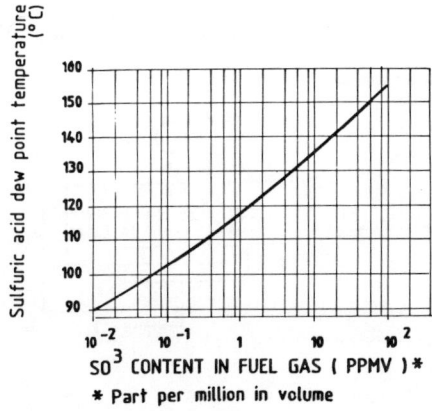

Figure 6-19. Dew points of H_2SO_4 in waste heat boilers. (Courtesy of J. R. Patterson.)

where t_d = the skin temperature of the tubes at the gas of their fins, °C
t_8 = the minimum economizer temperature (Figure 6-18), °C
t_7 = the minimum exhaust gas temperature (Figure 6-18), °C
24 = the estimated gas-side heat transfer coefficient, kcal/h-m²-°C
240 = the estimated liquid-side heat transfer coefficient, kcal/h-m²-°C

The minimum temperature in the economizer is set by the pressure in the deaerator (Figure 6-20).

The makeup boiler feedwater should be completely demineralized in ion exchangers; it should be filtered; and its pH should be adjusted to 6.5–8.5 through additions of hydrochloric acid and sodium hydroxide. Its conductivity should be 20 microsiemens/cm, with a maximum increase of 0.05. It should contain no more than 20 mg of silica per liter; and its CO_2 content should be less than detectable.

CHARACTERISTICS OF STEAM TURBINES

Steam turbines are described as "impulse" or "reaction" according to where the pressure drop occurs. If the total pressure drop occurs across the stationary set of blades, or nozzles, so that the flow through the moving rotor blades, or buckets, is at constant static pressure, the turbine is called an "impulse" turbine. If the pressure drop is divided equally between the stationary blades and the rotor blades, the turbine is called a "reaction" turbine. In both cases, a "stage" is considered to be one pair of stationary and rotating blades.

The action of steam turbines can be described in terms of velocity triangles (Figure 6-21) and the Bernoulli theorem of relative movement.

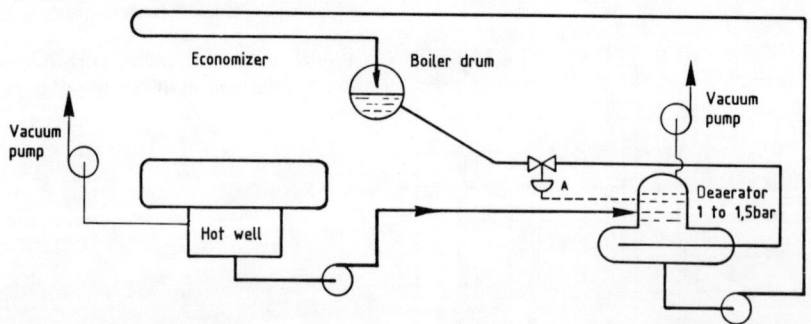

Figure 6-20. The deaerator sets the minimum temperature in the waste heat boiler.

Using the terms of Figure 6-21, the Bernoulli theorem can be expressed as follows:

$$\frac{W_2^2 - U_2^2}{2g} + \frac{p_2}{w} + z_2 = \frac{W_1^2 - U_1^2}{2g} + \frac{p_1}{w} + z_1 - \zeta_{12}$$

where ζ_{12} is the pressure drop between 1 and 2 in the wheel. Neglecting compressibility, V_{2m} and V_{1m} of Figure 6-21 are related as follows:

$$V_{1m} \cdot l_1 \cdot r_1 = V_{2m} \cdot l_2 \cdot r_2$$

The adiabatic expansion of gas through a stationary vane leads to work according to an energy balance as follows:

$$\mathfrak{I}_0 + J\,Q = \frac{V_2^2 - V_1^2}{2} + J\Delta i$$

where $\mathfrak{I}_0$ = work delivered at the shaft
Q = heat energy transferred
Δi = change in enthalpy
V = velocity
J = equivalence between heat and mechanical work

In the case of adiabatic flow through a fixed tuyere (Figure 6-22).

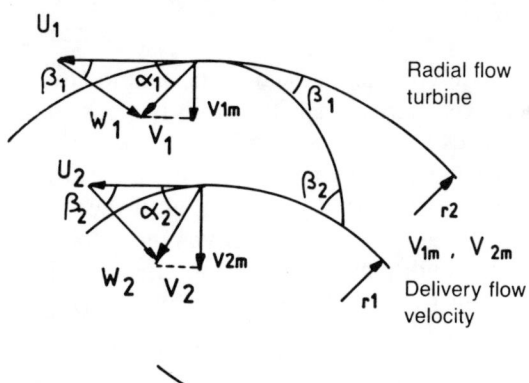

Figure 6-21. Velocity triangles in a radial impeller of a steam turbine.

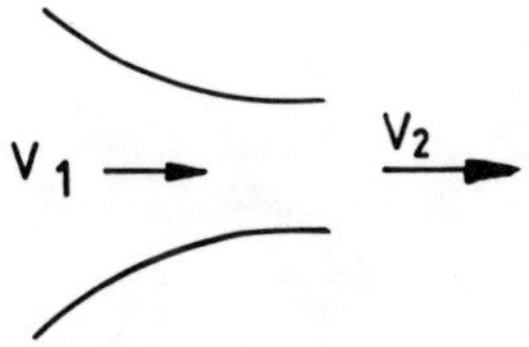

Figure 6-22. Flow velocities in a tuyere.

$$\frac{V_2^2 - V_1^2}{2} = J(i_1 - i_2)$$

$$\frac{V_2^2}{2} = \frac{V_1^2}{2} + q$$

where q is the loss of heat. This equation enables one to calculate the speeds of a fluid for a reversible adiabatic change at a given pressure based on a Mollier diagram for the fluid. For a perfect gas, the velocity expression can be related to pressure, p, and temperature, T, as follows:

$$\frac{V_2^2 - V_1^2}{2} = \frac{\gamma}{\gamma - 1} r T_1 \left[1 - \left(\frac{p_2}{p_1}\right)^{\frac{\gamma-1}{\gamma}}\right]$$

where γ = adiabatic coefficient of the gas
 r = perfect gas constant for 1 kg

These relations lead to calculations for the profile of a tuyere as follows:

$$I = \frac{S \cdot V}{v}$$

where S = the cross-sectional area, m²
 v = the specific volume, m³/kg
 I = the mass flow, kg/sec

If the fluid is liquid, its specific volume, v, is constant, and the tuyere converges as the fluid velocity, V, increases. If the fluid is a gas, however, its specific volume increases as its velocity increases, because the pressure decreases as the increasing velocity converts static pressure energy into kinetic velocity energy. This trend continues until the tuyere reaches a critical cross section, S_c, at which the velocity is the velocity of sound for that gas under those conditions. If subscript $_c$ denotes those conditions, relations for the critical cross section are as follows (Figure 6-23):

$$V_c = \sqrt{\gamma \, r \, T_c}$$

$$V_c = \sqrt{\gamma \, p_c \, v_c}$$

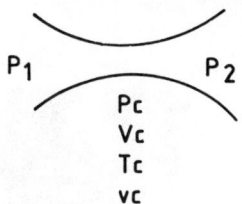

Figure 6-23. Characteristics of a tuyere (venturi).

$$p_c = p_1 \left(\frac{2}{\gamma + 1}\right)^{\frac{\gamma}{\gamma-1}}$$

And a neck is formed, after which the tuyere expands, if

$$\frac{p_2}{p_1} < \frac{p_c}{p_1}$$

According to Euler's theory, energy is recovered in a stage of a steam turbine according to the following equation by assuming an average flow element for narrow vanes and the integration of the flow elements for wide vanes:

$$I \cdot (U_1 V_1 \cos \alpha_1 - U_2 V_2 \cos \alpha_2)$$

where $U_1 = U_2$
I = mass flow

IMPULSE TURBINES

Within impulse turbines, the stationary vanes, or nozzles, convert the internal energy of the steam into kinetic energy of motion, and the moving vanes, or buckets, convert this kinetic energy into mechanical energy.

These turbines, which are axial-flow machines, consist of alternating fixed and rotating wheels (Figure 6-24). A large inlet volute feeds gas to expand through a ring of nozzles in a first stationary wheel. This ring is generally divided into four sections, and the nozzles are directed in an angle of 10°–12° from the plane of the wheel for better energy-conversion efficiency. From the nozzles the gas passes through vanes of a rotating wheel, which may be either one simple impeller (Rateau) or a double impeller with intermediate correcting vanes (Curtis impeller). There is no expansion through these intermediate correcting vanes and thus no change in temperature. Subsequently the gas expands through the nozzles

218 Drivers for Rotating Equipment

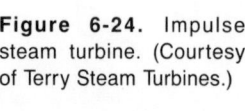

Figure 6-24. Impulse steam turbine. (Courtesy of Terry Steam Turbines.)

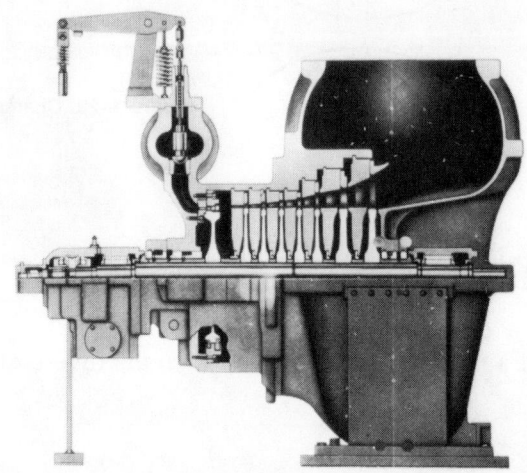

in a second fixed wheel before passing through the vanes of a second rotating wheel.

This sequence is repeated, with each pair of fixed and rotating wheels representing a separate stage. The fixed wheels have labyrinth seals on the shaft because they have different pressures on their two sides. Because there is no expansion and no change in pressure through the rotating vanes, there is no axial thrust. Also, because the flow is axial, $V_{2m} = V_{1m}$ (Figure 6-25). Because ζ_{12}, the pressure drop from 1 to 2 is zero, $p_2 = p_1$, and $W_2 = W_1$. And because the flow is axial, $U_2 = U_1$. This leads (Figure 6-25) to $\beta_2 = \beta_1$, and $\xi = U/V_1$. Only the change in kinetic energy produces work and the efficiency of the impeller, ρ, is as follows:

$$\rho = \frac{V_1^2 - V_2^2}{V_1^2} = 1 - \frac{V_2^2}{V_1^2}$$

The kinetic energies are as follows:

$$\frac{IV_1^2}{2}$$

and

$$\frac{IV_2^2}{2}$$

where I is again the mass flow.

Referring to Figure 6-25, it is apparent that the kinetic energy approaches its maximum as the angle α_2 of V_2 approaches $90°$ to the plane of the wheel, or as $\cos \alpha_2$ approaches 0. Furthermore, because the vane (or bucket) angles are constant, $\beta_2 = \beta_1$ and $W_2 = W_1$, and

$$V_2 \cos \alpha_2 + V_1 \cos \alpha_1 = 2U$$

from which (Figure 6-25)

$$\xi_{opt} = \frac{U}{V_1} = \frac{\cos \alpha_1}{2}$$

And at maximum efficiency

$$\rho_{max} = \cos^2 \alpha_1$$

This means that, by applying the Euler theorem, the hydraulic efficiency per stage, ρ, is the power recovered over the available energy, as follows:

$$\rho = \frac{I \cdot (U\, V_1 \cos \alpha_1 - U\, V_2 \cos \alpha_2)}{I \cdot \dfrac{V_1^2}{2}}$$

$$\rho = 2\, U \cdot \frac{V_1 \cos \alpha_1 - V_2 \cos \alpha_2}{V_1^2}$$

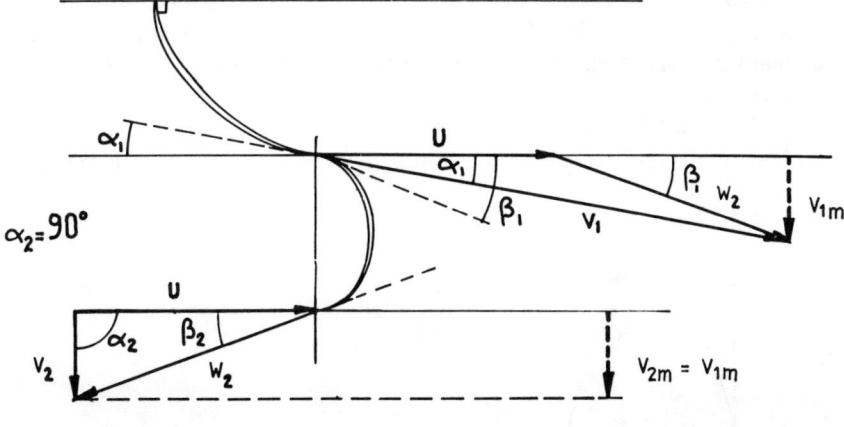

Figure 6-25. Velocity triangles in an axial impeller of a steam turbine.

Because $(V_1^2/2) = q$, the heat loss per unit mass per stage, the stage efficiency is equal to the efficiency per impeller. However, the terms of Figure 6-25 are also affected by friction. Velocity, V_1, is less than the theoretical velocity leaving the preceding nozzle, with a corresponding loss of heat energy, as:

$$V_{1th} = \sqrt{2q}: \quad V_1 = \varphi \sqrt{2q}$$

Similarly, velocity W_2 is lower than velocity W_1 ($W_2 = \psi W_1$); and the efficiency, ρ, is expressed as follows:

$$\rho = \frac{\varphi^2 (1 + \psi)}{2} 2U \cdot \frac{V_1 \cos \alpha_1 - V_2 \cos \alpha_2}{V_1^2}$$

$$\rho = \frac{\varphi^2 (1 + \psi)}{2} \cdot 4\,\xi\,(\cos \alpha_1 - \xi) \quad \text{per stage}$$

$$\rho = \frac{1 + \psi}{2} \underbrace{4\,\xi\,(\cos \alpha_1 - \xi)}_{\rho \text{ without friction}} \quad \text{per wheel}$$

where ξ and ψ are coefficients.

The optimum coefficient, ξ, equals $\cos \alpha_1/2$ as shown in Figure 6-26, but the exit velocity, V_2, corresponding to this optimum is not at 90° to the wheel, as seen for the theoretical, but is tilted slightly in the direction of rotation. For example, when $\alpha_1 = 10°$, coefficients φ and ψ are both 0.85, and the efficiency calculated by the preceding equations is $\rho = 0.67(0.98)^2 = 0.65$ per stage (0.88 per wheel). And the optimum coefficient $\xi_{opt} = 0.49$. The efficiency is also reduced by friction between the steam and the rotating wheels, more so as the wheels' surface gets larger compared to the nozzle surface.

In multistage turbines the inlets of the distributors of succeeding stages are arranged to recover the maximum part of the residual kinetic energy,

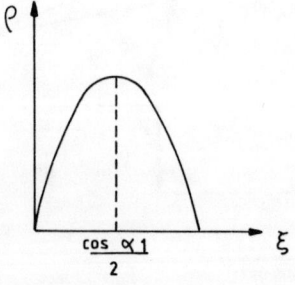

Figure 6-26. Efficiency vs. ξ coefficient.

$V_2^2/2$, from the preceding stage, which residual energy is lost in a single-stage turbine. Thus in multistage turbines the velocity V_1 leaving each of the successive distributors is as follows:

$$\frac{V_1^2}{2} = q + m\frac{V_2^2}{2}$$

where q = the heat loss per unit mass of steam
       m = the recovery (up to 1.0) of the preceding rotating vane

In order to achieve maximum recovery of the residual kinetic energy, the distributors must be as close as possible to the preceding rotating vanes, and the angle of the nozzle inlets must match the angle of V_2 leaving the preceding rotating vanes.

In one single stage the pressure and temperature of the steam are constant, thus the distributors' outlet velocity is constant and the rotating speed, N, varies as follows (Figure 6-27):

- Design speed N_0 for which efficiency is maximum and thrust is zero.
- Runaway at up to a maximum of about $2N_0$.
- According to the nominal torque, C_N.
- Upward from a startup torque equal to about $2C_N$.

When the impeller is doubled (two-stage impeller) with intermediate guiding vanes (Curtiss impeller), the distributor's kinetic energy output is recovered at constant pressure through two rings of moving vanes, which are supported on a single wheel and separated by a stator ring (Figure 6-28). The angles and velocities of the first and second rows of vanes are as shown in Figure 6-29 where $V_1' = V_2$ because the stator ring's vanes

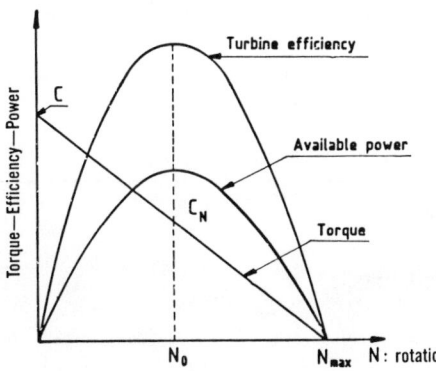

Figure 6-27. Variations of torque with speed in a steam turbine.

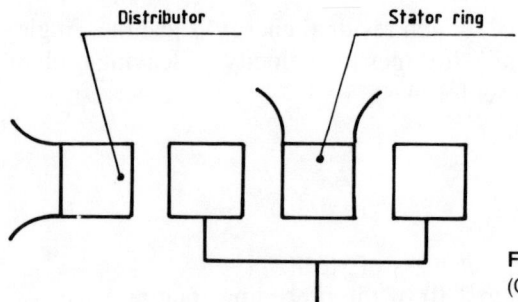

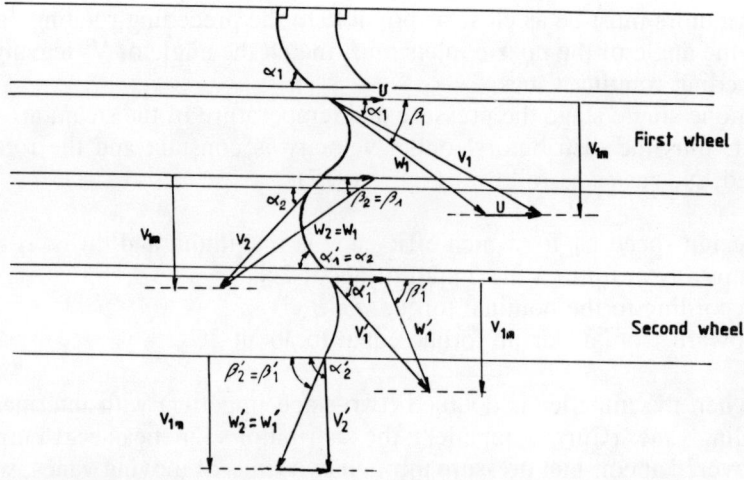

Figure 6-28. Principle of the double (Curtiss) impeller.

Figure 6-29. Velocity triangle in a two-stage impeller of a steam turbine.

serve only to direct the gas velocities without affecting either work or heat content. Assume that $\alpha_1' = \alpha_2$ for the entrance and exit of the first wheel, and the stator will have α_2 for both inlet and outlet. Also, $\beta_1 = \beta_2$ and $\beta_1' = \beta_2'$.

The efficiency, ρ, of a double impeller or of a stage with a distributor is calculated by the following:

$$\rho = \frac{I U (V_1 \cos \alpha_1 - V_2 \cos \alpha_2 + V'_1 \cos \alpha'_1 - V'_2 \cos \alpha'_2)}{I V_1^2/2}$$

where $IV_1^2/2$ = heat loss, q, per unit mass

and

$$V_1 \cos \alpha_1 + V_2 \cos \alpha_2 = 2U$$

$$V'_1 \cos \alpha'_1 + V'_2 \cos \alpha'_2 = 2U$$

$$V'_1 \cos \alpha'_1 = -V_2 \cos \alpha_2$$

The projected velocities are equal but in the opposite directions.

The preceding equations lead to ρ calculated as follows:

$$\rho = \frac{8U}{V_1^2}(V_1 \cos \alpha_1 - 2U)$$

If U/V_1 is called ξ, as with a single stage, the efficiency, ρ, becomes:

$$\rho = 8\xi(\cos \alpha_1 - 2\xi)$$

This efficiency is a maximum when ξ equals $\cos \alpha_1/4$, which corresponds for a single stage to the maximum exit velocity, V'_2, that is perpendicular to the plane of the wheel. The maximum efficiency is $\cos^2 \alpha_1$, as for a single stage.

The two-stage impeller affords four times the thermal load of a single impeller, as represented by V_1^2. Thus for a single-stage impeller:

$$2q_s = V_{1s}^2 = \frac{U^2}{\xi^2} = \frac{U^2}{\cos^2 \alpha_1}$$

and for a two-stage impeller:

$$2q_d = V_{1d}^2 = \frac{U^2}{\xi^2} = \frac{4U^2}{\cos^2 \alpha_1} = 4q_s$$

Consequently a two-stage impeller is equivalent to four single-stage impellers. However, when the losses of the distributor, rotating vanes, and guiding vanes are allowed for, the optimum coefficient ξ of the two-stage impeller is found to vary, while that of the single stage remains constant; and the two-stage impeller has a lower efficiency than the equivalent single-stage impellers.

If the friction in the correcting vanes is taken into account, coefficients φ and ψ are added to the formulas as follows:

$$V_1 = \varphi_1 \sqrt{2q} \quad \text{and} \quad V'_1 = \varphi_2 V_2$$

Drivers for Rotating Equipment

$W_2 = \psi_1 W_1$ and $W'_2 = \psi_2 W'_1$

The efficiency per stage becomes:

$$\rho = 2\,\varphi_1^2\,b\,\xi\left(\frac{a}{b}\cos\alpha_1 - \xi\right)$$

The efficiency per impeller becomes:

$$\rho = 2\,b\,\xi\left(\frac{a}{b}\cos\alpha_1 - \xi\right)$$

The maximum efficiency per stage becomes:

$$\rho_{max} = \varphi_1^2\,\frac{a^2}{2b}\cos^2\alpha_1$$

which corresponds to $\xi = \frac{a}{2b}\cos\alpha_1$.

And the maximum efficiency per impeller becomes:

$$\rho_{max} = \frac{a^2}{2b}\cos^2\alpha_1$$

where coefficients a and b are calculated from coefficients φ and ψ as follows:

$a = (1 + \psi_1) + (1 + \psi_2)\,\psi_1\varphi_2$

$b = (1 + \psi_1) + (1 + \psi_2)\,(1 + \psi_1)\,\varphi_2 + (1 + \psi_2)$

For example, when $\alpha_1 = 10°$, $\varphi_2 = \psi_1 = \psi_2 = 0.85$; $\xi_{max} = 0.237$ instead of the 0.49 based on no friction; and the maximum stage efficiency is $(0.85)^2 \times 0.76 \times (0.98)^2 = 0.53$, as compared to 0.66 for a single-stage impeller.

These calculations show the importance of friction between the steam and vanes, and thus why surfaces exposed to the steam should be smooth and why the shapes of the vanes should be well designed. For example, if the friction coefficients, φ_1, ψ_1, and ψ_2 can be brought to 0.90 through good vane design and manufacture, the efficiency per stage for a single-stage impeller is 0.75, and for a two-stage impeller 0.66.

Taking into account the inlet and outlet angles of the first distributor, it causes a drop in heat content that is three times greater than the drop across the second distributor.

Compared to the equivalent four single-stage impellers of equal diameter and speed, a two-stage impeller offers lower efficiency but also lower friction losses. Because the surface exposed by the two-stage impellers' rotating blades is larger than the surface of its nozzles, friction losses have a greater effect. All this means that the double impeller performs better at high steam pressure and low power, which corresponds to conditions at the first stage of a multistage turbine with less than 10,000 kW. This is where it is most used. Since a two-stage impeller corresponds to four single-stage impellers, it operates at a pressure equivalent to a fourth wheel, and it thus experiences less leakage. Compared to equivalent single-stage impellers, the two-stage impeller is less expensive and has a shorter shaft line, so that it is used for smaller low-power turbines with one two-stage impeller that experiences a faster temperature drop. By contrast, single-stage impellers have higher efficiencies as well as less erosion due to low inlet speeds.

Given the steam conditions and the available heat, the calculation of a steam turbine proceeds through five steps, as follows:

1. Choose the number and diameter of impellers and the type of impeller (two-stage or single-stage) for the first impeller.
2. Choose a coefficient, ξ, per wheel such that it approaches the optimum, $\cos \alpha_1/2$, when friction losses are taken into account.
3. Allow for partial recovery of friction losses, which are converted into available heat.
4. Calculate the distributor sections and the moving-vane angles.
5. Determine the steam leaks and allow for them.

Advantages of Impulse Turbines

Impulse turbines experience no axial thrust and no leaks across the rotating wheels, except that a slight thrust results from blade reaction during operation outside the design point. This slight thrust can be eliminated by perforations in the wheels to equalize the pressures on their two sides. Because there is no pressure difference across the rotating vanes, the turbine can easily be fed only half its normal steam to the first stage distributor without noticeably affecting its efficiency. This gives impulse turbines wide flexibility with respect to speed and power at unchanging efficiencies.

REACTION TURBINES

The steam's drop in heat content, as it passes through a stage of a reaction turbine occurs partly in passage through the distributor and partly in passage through the rotating vanes. If q_2 is the drop in heat content across the impeller and q_1 the drop across the distributor, the "degree of reaction" per stage, ϵ, is defined as follows:

$$\epsilon = q_2/(q_1 + q_2)$$

Accordingly, a pressure drop occurs across the rotating vanes as well as the fixed vanes.

The degree of reaction achieves an optimum at $\epsilon = 0.5$, with the fixed and rotating vanes identical, so that the contribution of each type might be exchanged for the other without affecting the efficiency. This supposes that the contour of both fixed and moving vanes causes the flowing velocity perpendicular to the wheels to remain constant. A minimum clearance between the rotating vanes and subsequent distributor enables recovery of virtually all the kinetic velocity, V_2, in the steam leaving the moving vane; the distributor's outlet velocity, V_1, is such that

$$\frac{V_1^2 th}{2} = \frac{V_2^2}{2} + q_1$$

Allowing for the coefficient of friction, φ, in the distributor vane leads to

$$V_1 = \varphi \sqrt{V_2^2 + 2 q_1}$$

With $q_1 = q_2$, the expression for the exit velocity of the moving vane is as follows:

$$W_2 = \varphi \sqrt{W_1^2 + 2 q_2}$$

Because the profiles fixed and moving vanes are identical (Figure 6-30), $\beta_2 = \alpha_1$, and $\alpha_2 = \beta_1$ and φ is the same. Applying the Euler theorem for one parallel stream (Figure 6-31) then leads to the following

$$\rho = \frac{I U (V_1 \cos \alpha_1 - V_2 \cos \alpha_2)}{I (q_1 + q_2)}$$

$$V_2 \cos \alpha_2 = W_2 \cos \beta_2 + U$$
$$= -V_1 \cos \alpha_1 + U$$

Steam Turbines—The Rankine Cycle 227

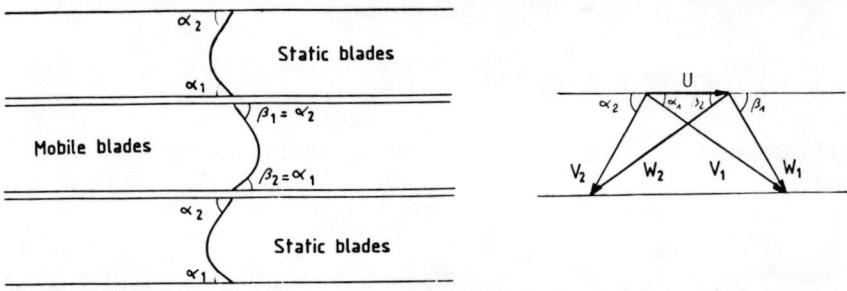

Figure 6-30. Velocity triangle in a stage of a reaction turbine.

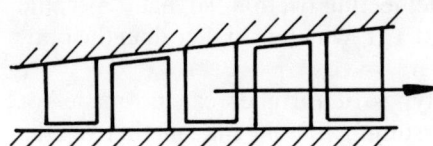

Figure 6-31. Steam parallel to the shaft.

$$2\,q_1 = \frac{V_1^2}{\varphi^2} - V_2^2 = \frac{V_1^2}{\varphi^2} - W_1^2$$

$$V_2 = W_1$$

$$2\,q_2 = \frac{W_2^2}{\varphi^2} - W_1^2 = \frac{V_1^2}{\varphi^2} - W_1^2$$

$$V_1 = W_2$$

$$q_1 + q_2 = \frac{V_1^2}{\varphi^2} - W_1^2$$

$$W_1^2 = V_1^2 + U^2 - 2\,U\,V_1\,\cos\alpha_1$$

$$\rho = \frac{U\,(2\,V_1\,\cos\alpha_1 - U)}{V_1^2\left(\dfrac{1}{\varphi^2} - 1\right) + U\,(2\,V_1\,\cos\alpha_1 - U)}$$

Assuming that the coefficient $\xi = U/V_1$, as for an impulse turbine, leads to the following:

$$\rho = \frac{\xi(2\cos\alpha_1 - \xi)}{\xi(2\cos\alpha_1 - \xi) + \left(\dfrac{1}{\varphi^2} - 1\right)}$$

and assuming the efficiency, ρ, becomes a maximum for $\xi = \cos\alpha_1$, so that:

$$\rho_{max} = \frac{\cos^2\alpha_1}{\cos^2\alpha_1 + \left(\dfrac{1}{\varphi^2} - 1\right)}$$

then this last equation shows that by neglecting friction, so that $\varphi = 1$, the ratios lead to an optimum efficiency of 1.0, which is higher than the optimum for an impulse turbine, where $\rho_{opt} = \cos\alpha_1^2$.

A comparison between different types of turbines can be made for given coefficients of friction, φ, by assuming that the inlet angle from the nozzles, α_1, is 10°, and calculating the efficiencies for one stage, as follows:

Type of Turbine	Efficiencies, ρ	
	Friction Coefficient, φ	
	0.85	0.90
Reaction	0.71	0.80
Impulse, simple impeller	0.66	0.66
Impulse, double impeller	0.53	0.75

For these values the coefficient, $\xi = \cos\alpha_1$, corresponds to an exit angle, $\beta_1 = 90°$, which in turn corresponds to an axial velocity, V_2, that is perpendicular to the plane of the impeller as it leaves the moving vane. If the degree of reaction is less than 0.5, β_1 is less than 90°; and if the degree of reaction is more than 0.5, β_1 is more than 90°. For the same mass velocity, U, and the same nozzle angle, α_1, the optimum coefficient, ξ, for a reaction impeller is twice that for an impulse turbine. The drop in heat content for a single impulse stage, q, is thus:

$$q = \frac{V_1^2}{2} = \frac{U^2}{2\xi_1^2}$$

And the similar drop in heat content for a reaction stage is:

$$q_1 + q_2 = V_1^2 - W_1^2 = \frac{U^2}{\xi_r^2} - W_1^2 = \frac{U^2}{4\xi_i^2} - W_1^2$$

Accordingly, if the residual outlet velocity of an impulse turbine is recovered, and if the effects of friction are neglected, the heat drop through one reaction stage is at the most only half of the heat drop through a comparable impulse stage. This means that reaction turbines need many more stages than impulse turbines, with larger shafts and drum-type rotors for rigidity. Because of the axial thrust experienced by the reaction impellers, such turbines must be divided in two and have balancing drums, which in turn bring on their leaks. Because of the pressure difference across the impellers and the greater diameters of the rotors, the leaks at the moving and fixed blades are greater. And because the division into sections of the inlet distributor would lead to more leaks, sectioning is not practical and partial injection is not possible with reaction turbines.

All in all this means that the superior efficiency of a reaction turbine is offset by leaks that increase with increasing steam pressure. Consequently reaction stages are usually used in compound turbines, in which the first stage is an impulse impeller. This shortens the rotor, reduces axial thrust, and affords greater flexibility by permitting partial injection, since its effects do not carry over so strongly on subsequent stages.

Calculating Wide Vanes

Although the average diameter to the fluid stream is satisfactory for calculating turbines with relatively narrow vanes, the variation with radius of fluid streams through wider vanes does have an effect on velocity triangles such as those of Figures 6-25, 6-29, and 6-30. The vane profiles must be altered to maintain the efficiency, with the best efficiency obtained by maintaining the moving vanes' exit velocity constant for all diameters; and this leads to a degree of reaction that increases with the diameter. A specific degree of reaction is given to the outer edge of the vane such that the degree of reaction at the base is zero. In order for the successive vanes to be parallel, each must furnish the same energy, and their exit velocities must be the same and axial.

According to the Euler theorem, the energy per impeller is as follows:

$$H = U_1 V_1 \cos \alpha_1 - U_1 V_2 \cos \alpha_2$$

$$U_1 = r_1 \omega$$

where ω = rotational speed
r_1 = radius

V_2 = constant

$\cos \alpha_2 = 0$

That is r V_{1U} is a constant for all impellers, and they are designed for a "constant flow." The maximum velocities occur at the base of the vane on entering from the distributor, and at the outer edge of the vane on leaving the vane.

If the velocities of the outside edges of the rotating vanes threaten to reach the speed of sound, it is possible to depart from a constant vane outlet velocity perpendicular to the rotating wheel. Although the pressure then varies from the inside to outside radii of the vane so that the flow stream is no longer in equilibrium, the leakage at the tip of the vane is reduced because the pressure is lower and the potential for erosion by any condensed water is reduced by the reduction in speed.

Reaction turbines are calculated like impulse turbines.

FEATURES OF STEAM TURBINE OPERATION

Steam turbines are distinguished by their ability to cope with condensation, sealing, steam flow, and safety.

Because of the difficulty of superheating steam to the point where it remains dry through all the stages of a turbine, there are turbines designed to handle condensation at the outlets of the distributors of the last stages. These turbines are called condensing turbines. The water droplets have lower velocities than the associated vapors, and thus different relative velocities in the moving vanes. This leads to shocks at the vane inlet and hence to a braking effect and erosion. These effects are more pronounced in reaction turbines because of the higher ξ coefficient and velocities. All in all, the amount of condensation should be restricted to 12%–13%, when reading down along a constant entropy line from the saturation point on a Mollier diagram (Figure 6-32). Also, the bad effects of condensation can be reduced by rings with drains at the outer edges of the impellers (Figure 6-32).

Sealing

Interstage seals are labyrinth seals. Reaction turbines have the clearance between the rotating vanes and the stator reduced by machining the vanes thinner at the edge. Also, the vanes can have grooves and ridges at the edges so as to form labyrinth seals with the stator.

The seals to the outside at the shaft are labyrinth with intermediate bleeds, or at moderate peripheral speeds and temperatures, carbon-ring mechanical seals (Figure 6-33).

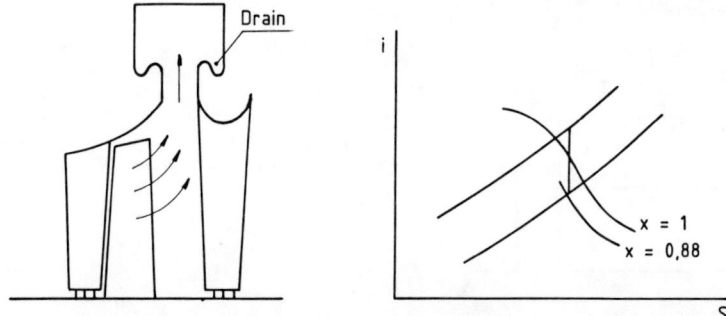

Figure 6-32. Condensation during expansion within a steam turbine.

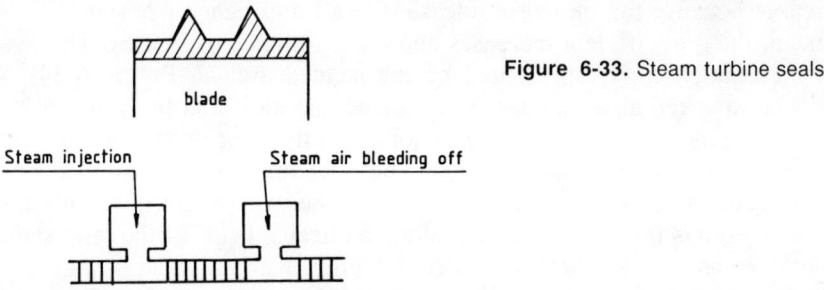

Figure 6-33. Steam turbine seals.

Whenever it discharges to a vacuum condenser a turbine's latter stages will operate under constant vacuum, and the entire turbine may be under vacuum when shut down. To prevent air from entering, steam is injected into the seals at slightly above atmospheric pressure, and a mix of steam and air are drawn off at an intermediate bleed by a vacuum pump or an ejector with small condenser.

Regulating Steam Flow

The speed and thus the power of steam turbines is controlled either by a throttle valve with streamlined seats or a partial injection of steam.

The streamlined control valve is designed so that the steam can expand at constant velocity, and thus at constant enthalpy. Since there is no work, $\mathfrak{J}_e$, or exchange of heat, Q, as the steam passes through the valve, the change in velocity head, $\Delta V^2/2$, equals the change in enthalpy, Δi; and since the velocity is constant, both changes are zero.

$$\mathfrak{J}_e + Q = \Delta \frac{V^2}{2} + \Delta i$$

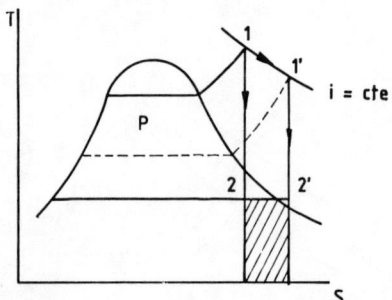

Figure 6-34. Heat loss in the condenser.

However, there is an increase in entropy as the conditions of the steam are changed to a lower pressure, a slightly lower temperature, and a greater degree of superheat. At constant speed, the inlet velocity decreases because the enthalpy release overall and hence per stage is reduced, the ξ coefficient increases and the efficiency decreases. The loss at the condenser is represented by the shaded area in Figure 6-34. A change in speed also reduces the efficiency as indicated in Figure 6-28.

Control by partial injection does not affect the heat release and has little effect on the stage efficiencies. The inlet distributor is divided into four sections fed successively by a streamlined valve. Because only the first stage has this divided distributor, the heat release of the first stage increases as the pressure to the first impeller decreases due to a greater drop through the distributor. Consequently the heat release, and thus the power, of the first stage increases as that of the other stages decreases. The efficiencies vary accordingly.

Safety

Steam turbines are equipped with an independent safety relief valve upstream of the inlet control valve. This safety valve is closed by over speed, lack of oil pressure, vibration, or excessive vacuum at the condenser.

Chapter 7
Couplings

A coupling is the device that makes the mechanical connection between a power-delivering shaft projecting from an engine or motor and a power-receiving shaft projecting from a pump, compressor, or other rotating machine. Couplings can be described as direct, speed-reducing or speed-increasing, or as variable speed.

DIRECT COUPLINGS

Direct couplings deliver 100% of the prime mover's turns to the driven machines. They can be either rigid or flexible, but in either case their use implies aligning the shafts of the prime mover and driven machine.

Rigid Direct Couplings

These generally consist of two plates, each fastened to its shaft and carrying bolt holes that match the bolt holes in the other, so that the two plates can be bolted together. Such couplings are suited to large loads, but they require perfect alignment between the two shafts, and they are not ideally suited for high rotating speeds.

Flexible Direct Couplings

These do not require such perfect shaft alignment as rigid couplings. They can vary from flexible springs that connect matching splines on the two shafts (Figure 7-1), through gears (Figures 7-2, 7-3, and 7-4), to flexible plates and diaphragms of metal or rubber (Figures 7-5, 7-6, and 7-7). Depending on the type, they can absorb some shock and reduce vibration from one shaft to the other.

The single-geared couplings (Figure 7-2) are usually continuously lubricated. Double-geared couplings (Figure 7-3) are lubricated either continuously or with oil under pressure in a casing. They are used for high speeds (5,000–15,000 rpm) and high powers up to 100,000 kW. The teeth of the gears are slightly rounded so that the couplings can tolerate a slight disalignment such as might occur from thermal expansion. The

234 Drivers for Rotating Equipment

Figure 7-1. Citroen flexible coupling.

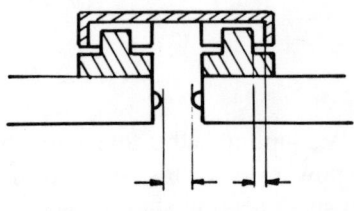

Figure 7-2. Gap limiting shaft displacement in a gear coupling.

Figure 7-3. Amerigear flexible couplings. (Courtesy of Zurn Industries, Inc.)

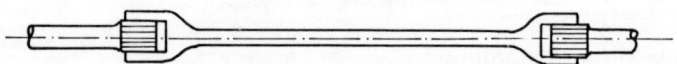

Figure 7-4. Geared coupling with spool.

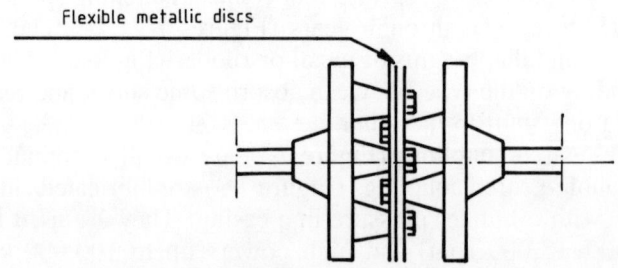

Figure 7-5. Principle of disc-type flexible couplings.

manufacture of gear couplings is exacting. Gear couplings with torsional shafts (Figure 7-4) are suitable for speeds up to 25,000 rpm and powers of 1,000 kW.

Other couplings achieve flexibility through the use of rubber to transmit the load. The rubber may be in the form of blocks extending from grooves in one gear face to matching grooves in another (Figure 7-8), or it may be in the form of a rubber tire whose two rims are gripped in the separate coupling faces (Figure 7-7).

Flexible thin metal plates in multiple layers are the principle on which still another type of flexible direct coupling is designed. Two shaft faces are joined by means of alternating bolts (Figure 7-5). Couplings with two flexible thin metal plates in between a rigid spool (Figure 7-6) are good for high speeds (5,000–15,000 rpm) and high power up to 100,000 kW, and they can accept a disalignment up to .5°.

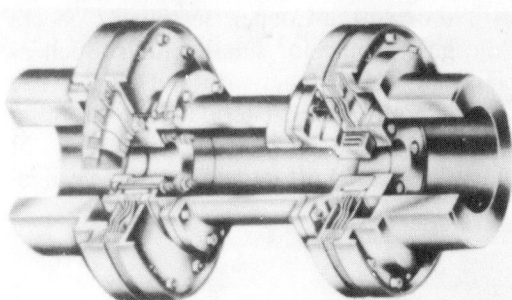

Figure 7-6. Ameriflex^R flexible diaphragm couplings. (Courtesy of Zurn Industries, Inc.)

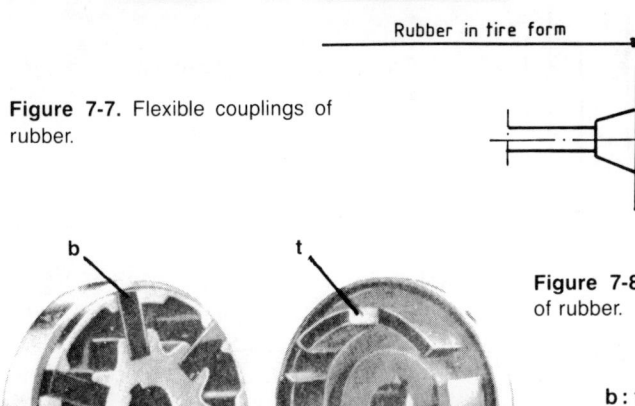

Figure 7-7. Flexible couplings of rubber.

Figure 7-8. Coupling with blocks of rubber.

b: flexible blocks
t : tenon
F: female part
M: male part

Aligning the Shafts

The shafts should be aligned as perfectly as possible. The difference in distance between opposite sides of the shaft flanges should be less than 0.05 millimeters, when measured from the sides of the flange (1, in Figure 7-9), and less than 0.03 millimeters, when measured from the flange faces (2, in Figure 7-9). In the latter measurement, lateral clearance in the bearings must be compensated for by pushing the shafts. The lateral clearance should be sufficient so as not to be canceled by lateral movement of the shafts during rotation. Measurements should be made at the normal operating temperatures. If they are not, they should be corrected for thermal expansion. When aligning two machines, only one machine at a time should be moved.

When aligning compressors and gas turbines, the machine first positioned will be either the compressor for large units, such as an Ingersoll Rand GT 50 or 60, a Clark DJ 270 or 240, a Cooper Bessemer Coberra 2,000, 3,000, and 6,000, or the gas turbine for smaller units, such as Solar Saturn, Centaur, Mars, Hispano THM, Ingersoll Rand GT 22. The base plate of such machines has at its four corners lateral bolts for forcing the machine into proper horizontal position, as well as shims made of layers of thin metal sheets that can be peeled off to get the exact vertical position (Figures 7-10, 7-11, and 7-12).

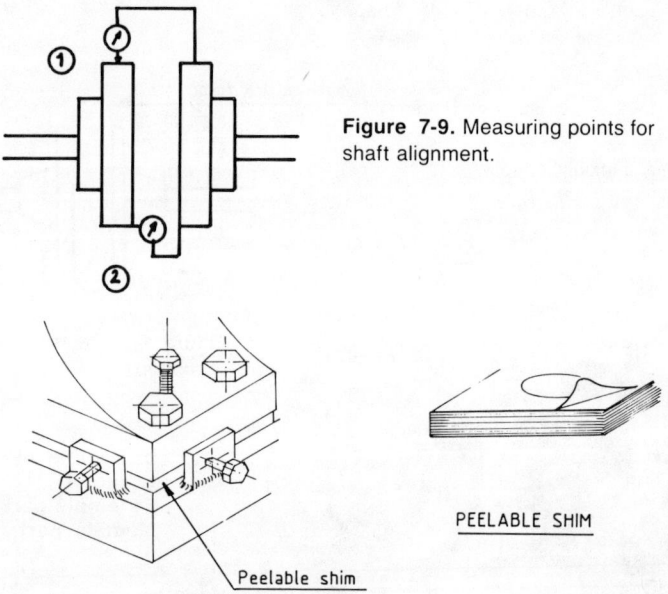

Figure 7-9. Measuring points for shaft alignment.

Figure 7-10. Shims and bolts for aligning heavy machines.

Couplings 237

Figure 7-11. Horizontal corner bolts for aligning a compressor.

Figure 7-12. Gas turbine displacement by action on the gas turbine mounting.

SPEED-REDUCING AND SPEED-INCREASING COUPLINGS

These couplings transmit power at a fixed ratio of rotating speeds between the shafts of the prime mover and driven machine. They always use gears, which are classified according to whether all the gears are fixed in one position at the axis, or whether some of the gears can move around within the housing. The first group includes spur, helical, and bevelled gears; the second group includes epicyclic gears.

Fixed-Gear Couplings

These couplings can handle speed changes across one or two stages, as shown:

Number of Stages	Maximum Speed Ratio	Mechanical Efficiency Guaranteed	Actual
1	10	98%	99%
2	30	96%	97%

Epicyclic-Gear Couplings

Epicyclic gears all include four principal parts, as follows:

1. A central smaller pinion gear, called a "sun gear," directly in line with a drive shaft.
2. A group of gears whose teeth mesh with those of the sun gear and whose shafts are held in a plate, so that they are free to move around the sun pinion gear. These gears are called "planet gears," and the plate the "planet carrier."
3. The plate just mentioned.
4. An encircling gear with inside teeth that mesh with the teeth of the planet gears. This encircling gear is called the "orbit gear."

Epicyclic gears are divided into two classes, depending on how the load is transmitted; if it goes from the pinion sun gear to the planetary gears and hence to the planet carrier, which is attached (Figure 7-13) to the mating shaft, the gear train is called "planetary." If it goes from the pinion sun gear to the planetary gears, which rotate the orbit gear, which is attached to the mating shaft, it is called a "star" gear (Figure 7-14). A third type rarely used has a fixed sun gear, with one shaft attached to the planet carrier and the mating shaft attached to the orbit gear (Figure 7-15).

Planetary gears are good for high-torque applications at speed ratios of from 3/1 to 12/1. Star gears are good for high-speed applications at speed ratios of from 2/1 to 11/1. The fixed pinion gear is good for low speed applications at speed ratios of from 1.1/1 to 1.7/1. Also, epicyclic gears can be used in more than one stage; and two stages of epicyclic gears with speed ratios of 30/1 have been used to drive a 330 rpm reciprocating compressor with a turbine turning at 10,000 rpm.

Epicyclic gears have several advantages that may not be immediately apparent. Because the pinion sun gear is supported partly by the planetary gears, its bearing can be eliminated and its position maintained by the machine's bearings. Because the load is distributed evenly over 3 to 7 planetary gears, the gear teeth can be weaker and shorter. Thus the size of the epicyclic gear train is much smaller and about half the weight of a train of equivalent fixed gears. Furthermore the mating shafts are coaxial; the operation is smooth and reliable; and the mechanical efficiency is excellent, on the order of 98.5%–99.25% for single stage and 97.75%–98.25% for two stages; due to the absence of support bearings. Epicyclic power transmission can be as high as 50,000 kW.

Rubber Drive Belts

These are limited to 1,000 kW power transmission. Their efficiencies are 96%, 95%, and 94% at full, three-quarters, and half loads, respectively.

Couplings   239

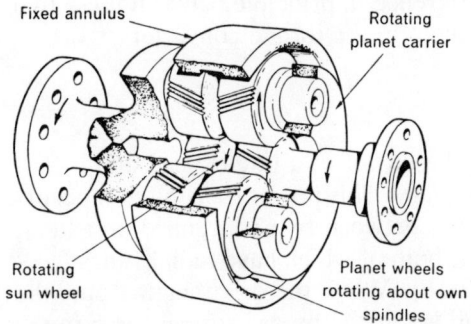

Figure 7-13. Planetary epicyclic gear. (Courtesy of NEI-APE Ltd.—Allen Gears.)

Figure 7-14. Star epicyclic gear. (Courtesy of NEI-APE Ltd.—Allen Gears.)

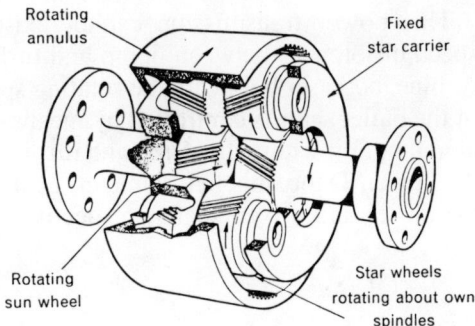

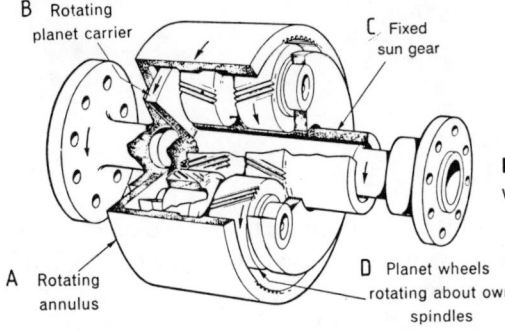

Figure 7-15. An epicyclic gear train with fixed sun gear.

VARIABLE-SPEED COUPLINGS

Except for belt-driven variable speed couplings, which are limited to 80 kW and 3.6/1 speed ratios at efficiencies of 96%, 95%, and 94% at full, three-quarters, and half loads, respectively, variable speed couplings are hydraulic, with power transmitted by circulating oil. The hydraulic couplings are further differentiated as "fluid couplings" and "hydraulic torque converters," the difference in principle between these two being the presence of a reaction stage in the torque converters.

Hydraulic Couplings

A hydraulic coupling transmits the power of a fixed-speed drive shaft to an outlet shaft whose speed may vary. This power transmission occurs through the medium of oil confined between two concentric impellers, one a pump and the other a turbine, both of which have radial vanes face-to-face with each other (Figure 7-16). When the drive engine puts the pump impeller into motion, the oil is carried by centrifugal force to the pump's periphery, from whence it enters the turbine impeller with the force necessary to put the turbine into rotation; and power is transmitted.

This power transmission cannot exist without slippage, and hence speed differences between pump and turbine. The power exerted on the turbine increases both with the rotating speed and the slippage. The speed of the outlet shaft is controlled by means of a bleed bypass that varies the amount of oil within the pump and turbine impellers. If N is the speed, W the power, D the impeller diameter, M the torque, and subscripts 1 and 2 the driver and driven shafts, respectively, these are related as follows:

$$M_2 = f\, N_1^2\, D^5$$

$$M_1 = M_2$$

$$W_1 = M_1\, N_1$$

$$W_2 = M_2\, N_2$$

$$(W_2/W_1) = (N_2/N_1)$$

The efficiency is on the order of 97.5%, because a slippage of 2%–3% is necessary. Efficiency decreases with a decrease in the speed of the outlet shaft, so that hydraulic couplings are more suited for driving machines whose power declines with declining speed, such as centrifugal compressors for which power is proportional to the cube of the speed. With such machines, low efficiencies occur only at low power transmission. By contrast, this type of coupling is not at all suited for driving low-speed volumetric machines.

For operations at up to 12,000 kW, the speed ratios for hydraulic couplings range between 20% and 100% for machines, such as centrifugal compressors, which have parabolic characteristic curves, or between

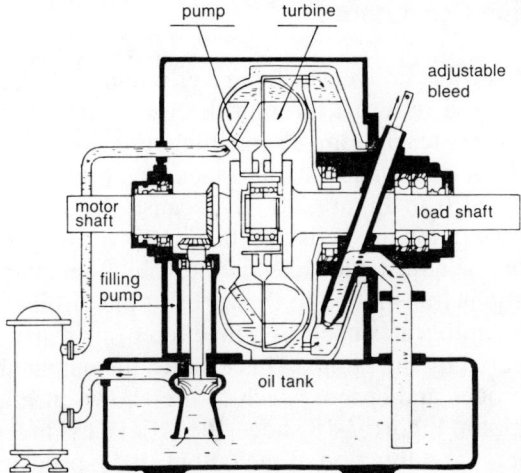

Figure 7-16. Operating principle of a hydraulic coupling. (Courtesy of Alsthom-Atlantique-Neyrtec.)

30% and 100% for constant torque machines, such as volumetric compressors.

A hydraulic coupling powered by an electric motor offers the following advantages:

- The motor can start up at no load, because the driven machine is only engaged when the motor attains its maximum torque.
- The acceleration of the driven machine is done entirely by the coupling and can be as gradual as desired.
- It is possible to engage and disengage the driven machine, when otherwise the disengagement of a compressor, for example, might lead to a runaway turbine or motor.
- The electric startup equipment is greatly simplified; and it is even possible to start up on full power.
- It is possible to use simple and rugged squirrel-cage motors.
- Explosion-proofing is easy.

The hydraulic coupling functions analogously to a sliding rheostat. However, the fixed losses of hydraulic couplings are due partly to friction losses at the vanes and partly to the power needed to get the oil circulating on startup; and the rheostats experience smaller friction losses and no loss analogous to the oil circulation. On the other hand, automatic control is easier and more reliable with a hydraulic coupling; and a sliding rheostat requires an asynchronous motor.

Hydraulic Torque Converters

In a hydraulic torque converter the driver pump acts like a centrifugal pump to force liquid outward through the vanes of a turbine impeller, into recycle passages resembling a centrifugal compressor's diffuser with fixed vanes and volute with adjustable vanes, and back to the pump again (items 1, 3, 5, and 6 in Figure 7-17). Because the speeds of the input drive shaft and the pump are constant, the power developed by the pump is determined by the adjustable vanes in the volute; and the power developed by the pump is transmitted to the vanes of the turbine. The speed of the output shaft equilibrates itself as a function of its resisting torque and the power delivered by the pump. This enables the output shaft to meet a full range of torques and speeds, such as are shown in Figure 7-18.

The bottom plot in Figure 7-18 shows the efficiency, in terms of output over input powers, as a function of the output shaft speed and vane openings. As long as the openings are above 60% of wide open, the efficiency at speeds of 60%–110% remains at 80%–85%. The top two plots show the range of torques and powers encompassed by these conditions. They show that the hydraulic torque converter is well suited for driving machines with which the torque can vary independently of the speed—machines such as reciprocating compressors. Otherwise, the hydraulic torque converter offers the same advantages as the hydraulic coupling, such as no-load startups, simplified electrical equipment, and so forth. However, this type of coupling should be equipped with a clutch to prevent a runaway of the driven machine in the event of a loss of resisting torque that happens faster than the adjustable vanes can be closed. In such cases, the hydraulic torque converter can act like a speed-increasing coupling.

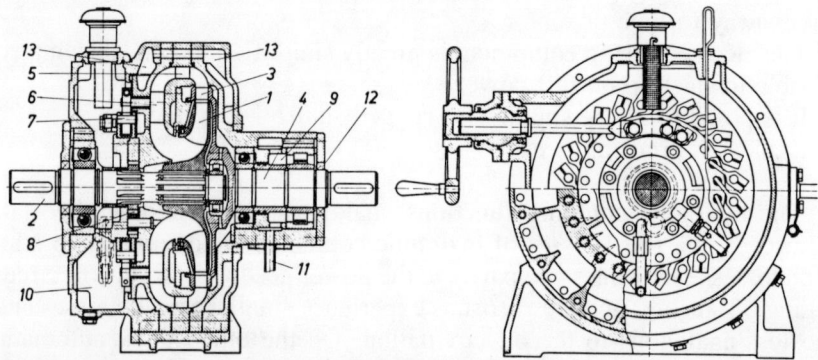

Figure 7-17. Principal operating parts of a hydraulic torque converter: 1-pump; 2-primary shaft; 3-turbine; 4-load shaft; 5-fixed vanes; 6-mobile vanes; 7-control of the fixed-vanes position; 8-gear pump; 9, 10-oil tank; 11-oil pipes; 12-seals; 13-cooling water. (Courtesy of Voith.)

Couplings 243

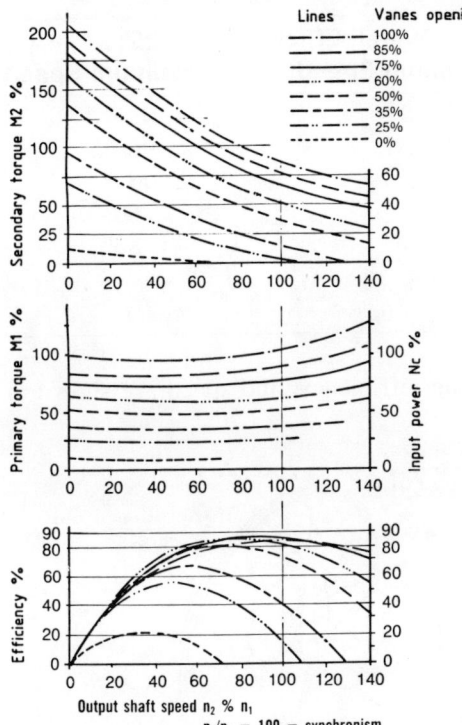

Figure 7-18. Operating curves for a hydraulic torque converter. (Courtesy of Voith.)

Variable-Speed Epicyclic Gear Couplings

Referring to the epicyclic gear couplings in Figures 7-13 and 7-14, it can be seen that the orbit gear can be rotated at various speeds in relation to the sun gear through the intermediary of the planet gears, and that such variable relative rotation can be used to make the epicyclic gear train a variable coupling as well as a speed-changing coupling. Figure 7-19 shows one system for accomplishing this by means of a closed hydraulic circuit and two hydraulic reciprocating pumps, one geared to the sun gear and the other to the orbit gear. In Figure 7-19 the output shaft is turned by the planet gears, whose speed is a function of the speeds of the sun gear and the orbit gear, which speeds can be varied relative to each other by means of the hydraulic circuit; the orbit gear can be stationary, rotated in a sense opposite to the sun gear, or rotated at varying speeds in either direction.

This system is well suited to automatic control. It affords transmitting a constant torque, or a variable torque, at only a slight sacrifice of efficiency; and it is suited for all power levels. The efficiencies at various speeds are as follows:

244 Drivers for Rotating Equipment

Amplitude of Speed Variation, %	Overall Efficiency ±0.25%	
	At Maximum Speed	At Minimum Speed
10	98.0	98.0
20	97.25	96.75
30	96.5	95.5
60	92	60[a]
		86[b]

[a] With outlet torque varying as the cube of the speed.
[b] With constant outlet torque.

The relations between outlet torque, efficiency, and speed are shown in Figure 7-20.

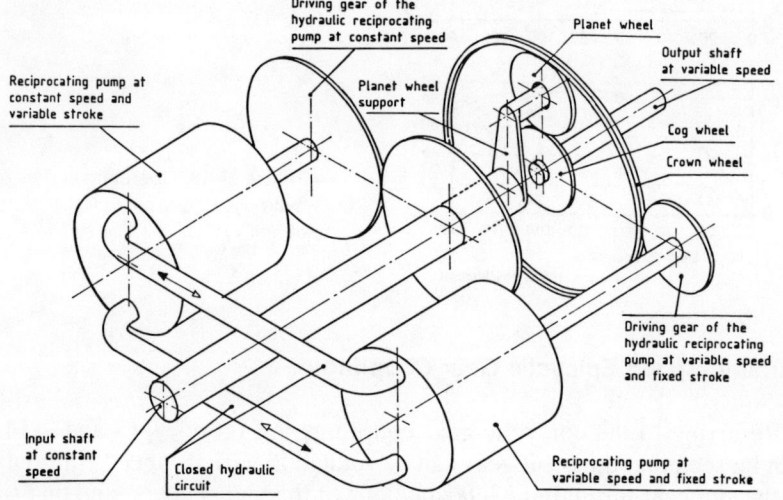

Figure 7-19. Operating principle of the variable-speed epicyclic gear. (Courtesy of NEI-APE Ltd.—Allen Gears.)

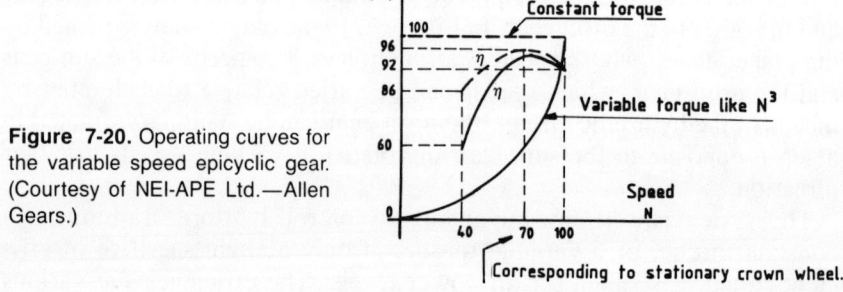

Figure 7-20. Operating curves for the variable speed epicyclic gear. (Courtesy of NEI-APE Ltd.—Allen Gears.)

Index

Air filters
 classes of, 36, 38, 40
 specifications, 36, 38
Air gap, 5, 6
Air-gas ratio, 57–58, 101
Alternator, 75
Ambient conditions, effect on gas engines, 98
Ambient temperature, effect of, 91
Ammonia injection, 72–73
Anti-ice systems, gas turbine, 182–183
Apparent power, 1
Applications
 free-piston gas generators, 187
 impulse steam turbines, 225
Ash in oil, 64
Auto-ignition (pre-ignition), 64

Baffle filters, 44
Bag filters, 40, 43–44
Beau de Rochas cycle, 16, 19–20
Bernoulli theorem, 214–215
Bevelled gears, 237
Boiler feedwater, 189, 195
 makeup, 213–214
 pumps, 194
Brayton cycle, 109–110, 157
Buffer winding, 5

Calculations
 for a gas turbine, 117–119
 inlet and outlet conditions, 176–177
 steam turbine, 212–213
  steps of, 225, 230
 waste heat boiler, 207–212
Carbon monoxide emissions, 29–31
Carnot cycle, 18
Catalyst efficiency (NO_x reduction), 72

Catalytic converters, 69–73
Catalytic NO_x reduction, 72
Characteristic curves, gas turbine, 148, 160–164
Characteristics of steam turbines, 214–230
Cheng cycle, 157
Closed cycle, gas turbine, 110–112
Combined cycle, 16, 22–23
 compressors with, 137–138
 design for gas turbines, 140
 gas-steam, 142–312
Combustion efficiency, boiler, 190
Combustion temperature, maximum, 24
Control
 of gas turbines, 181
  combined cycles, 138–139
 of steam drums, 195
 of waste heat power generators, 204–206
Corrections for operating conditions of gas turbines, 166–181
Couplings, 233–244
 speed reducing and speed increasing, 237–244
 variable speed, 239–244
Curtain filters, 40–42
Current frequency, 1
Curtis impeller, 217, 221–222
Cycloconverter speed regulator, 9–10
Cyclone filters, 40, 43
Cylinder pressure monitoring, 104

Deaerator, 194
Delivered power, 1
Design criteria for gas turbines, 142–149
Detonation, 65, 67
 gas engine, 60–64
 dew points, H_2SO_4, 213

Diesel cycle, 10–22
Direct couplings, 233–236
  flexible, 233–235
  rigid, 233
Double cage rotor, 3
Downflow draft, 203–204
Dual combustion, 66
Dual fuel engines, 73
Dual-pressure steam systems, 198–202
  once through, 200–204
Dust
  contents in ambient air, 38
  fine, 35
  particles in ambient air, 39
  particulate, 36
  types of and effects of, 37

Efficiency
  Beau de Rochas cycle, 22
  combined-cycle gas turbines, 142
  combustion, 26–27
  cycle, 22
  diesel, 22
  gas turbines
    thermodynamics, 23
    with and without heat recovery, 133
  indicated, 27
  mechanical, 27
  nonsynchronous motors, typical, 4
  reciprocating internal combustion
      engines, 27
  shape, for thermodynamic cycles,
      25–26
  steam turbine
    impulse and reaction, 228
    systems, 188–189
  synchronous motors, typical, 6

Electric motors, 1–15
Electronic ignition, 73–77
Element-type filters, 40–45
Energy balance
  around a gas engine, 59
  with a steam cycle, 134–135
Energy consumption, gas engine, 97
Engine cycles with and without
    turbocharging, 88
Enthalpy of air, 124–129
Epicyclic gear couplings, 237–238
  variable speed, 243–244
Euler's theorem, 217, 219

Exhaust temperature, gas turbine, 164
Expansion turbine, 87

Faraday's law, 3
Field excitation, 1
Filters
  baffle, 44
  bag, 40, 43–44
  double, 43
  recommended, 46–47
  reverse flow, 44
  self cleaning, 46
Final combustion temperature, 62–63,
    65–66
Final compression temperature, 62, 65,
    69
Finned tubes, 200
First- and second-generation gas
    turbines, 146–149
Fixed-speed alternating current motors,
    2–6
Flame temperatures, air-methane, 30
Flexible disc couplings, 234–235
Flow control, steam turbine, 231–232
Fluid couplings, 239
Four-cycle gas engines, 50
Free-piston gas generators, 87, 184–187
Free-turbine gas turbines, 143–144
Friction coefficients, steam turbine, 220,
    223–224, 226, 228
Fuel consumption, calculation for gas
    turbines, 177–178
Fuel corrections for gas turbines, 171
Fuel quality, gas engine, 101

Gas engines
  ambient conditions, 98
  cylinders, 50
  detonation, 60–64
  dual-fuel oil ignited, 50
  four-cycle, 50
  fuel quality, 101
  ignition, 58–59, 73–85
  maintenance, 105–107, 153–154
  maximum pressures, 106
  maximum temperatures, 119–120
  mean effective pressure, 55–56, 93,
      99–101
  NO_x emissions, 65–73

oil consumption, 98
on-stream ratio, 50
operating characteristics, 96–107
piston displacement, 55
pollution, 58
power, 50, 55
pressure-volume diagrams, 52–53
scavenging loop, 53–55
spark advance, 59, 98–102
spark ignited, 50
specific consumption, 97
speed regulation, 96
supercharging, 53–55
thermal efficiency, 50, 56–57
timing, 51–52
two-cycle, 51
volumetric ratio, 56–57
Gas turbines, 108ff
   air compressors, 149–156
   anti-ice systems, 182–183
   blades, 152
   calculations, 117–119
   characteristic curves, 148, 160–164
   characteristics, 108–109, 160–166
   closed cycle, 110–112
   combined cycle design, 140
   combustion chambers, 111, 150–151
   compressors, 110, 114, 118, 120, 143–144, 147, 149
   control, 181
   corrections for operating conditions, 166–181
   cycle, 16, 23–25
   cycle diagrams, 113
   design criteria, 142–152
   dust, 36
   efficiency, 23, 133
   exhaust gas cooler, 110, 122, 143
   exhaust temperature, 164
   expansion turbines, 110, 114, 118, 120, 143–144, 149, 151–152
   first- and second-generation, 146–149
   free turbines, 143–144
   fuel consumption, 171, 177–178
   gas generators, 144, 147, 160–161, 183
   gas heater, 110, 143, 147, 150–151
   heat balance, 164
   heat exchangers, 122–123
   injections, 155–157
   inlet and outlet conditions, 176–177
   life expectancy, 120, 162–163

   low-boiling fluids, 136
   oil consumption, 166
   open-cycle, 111
   partial loads operation, 139–140
   pollution, 34–35
   power
      available, 177
      variation, 110
   pressure corrections, 166, 167, 169, 170, 174–175
   shaft lines, 143–144
   specification curve, 162–163
   speed variation, 144, 165
   startup, 166, 181
   steam injections, 156–157
   temperature corrections, 173
   thermodynamic efficiency, 114
   water injections, 155, 157
   work at the shaft, 115
Geared couplings, 233–234
Graetz bridge, 9–11, 13
   voltage from, 10

Heat balance, gas turbine, 164
Heat engines, characteristics of, 16–49
   kinds of, 16
High altitude engines, 90
Helical gears, 237
High-pressure boiler, 188
Hot well, 194
Humidity corrections, 171, 174
Hydraulic couplings, 240–241
Hydraulic dampener, 96
Hydraulic leg, 194
Hydraulic torque converters, 240–243
Hyposynchronous cascade, 7–9

Ignition
   advance, 61
   control, 98–101
   delay, 68
   explosion proof, 78–82
   gas engine, 58–59, 73–85
   of free-piston gas generators, 185–186
   shielded systems, 80–83
   timing, 60–61, 64
   voltage, 73–75
Impedence, 1
Injections, gas turbine, 155
Intake air filters, 35–46
Intake silencers, 95–98

Integrated gas engines and compressors, 94–95
ISO conditions, 109

Lambda probe, 70–71
Life expectancy
  gas turbine, 120, 162–163
  heat engines versus dust, 36
Liquid carryover, 203
Low-boiling fluids in gas turbines cycles, 136

Mach number, 167
Magnetic poles, 1
Maintenance, gas engine, 105–107
Materials for gas engine parts, 153–154
Maximum pressures, typical gas engine, 106
Maximum temperatures, typical gas engine, 119–120
Mean effective pressure, gas engine, 55–56, 93, 99–101
  control of, 102–105
Mist separators, 41
Multistage combustion, 158

NEMA conditions, 109
Natural commutation, 12, 14
NBN (normal butane number) index, 61
NO_x emissions, 29–31
  gas engine, 65–73
  in gas turbine exhaust, 152, 155–160
  versus specific fuel consumption, 58–59, 66
  U.S. limits, 32–33
NO_x formation, 29, 31
NO_x reaction inhibition, 159
NO_x reduction, 155–157, 159
Nonsynchronous motors, 2–4, 6–12

Octane number, 60
  of light hydrocarbons, 62
Oil-bath filters, 40–41
Oil consumption
  gas engine, 98
  gas turbine, 166
Oil wedge, 85
On-stream ratio, of gas engines, 50
Open-cycle gas turbine, 111
Open cycle with and without heat recovery, 129–132

Operating characteristics
  of dual pressure boilers, 200
  of free-piston gas generators, 185–186
  of gas engines, 96–107
  of gas turbines, 160–166
  of steam turbines, 230–232
  of waste heat boilers, 195–196
Operating gas turbines at partial loads, 139–140
Operating principle, free piston gas generators, 184–185
Orbit gear, 238
Oxygen in exhaust gas, 70–71

Phase angle, of voltage, 2
Pinch point, 198, 202, 208–209, 211
Pinion gear, 238
Piston displacement, gas engine, 55
Piston-liner speeds, 93
Piston rings, 85–87
Planet carrier, 238
Planet gears, 238–239
Pollution
  gas engine, 58
  gas turbine, 34–35
  heat engine, 28–35
  reduction methods, 34–35
Polluting agents
  calculation of, 31–32
  equivalences of, 31
Power available, gas turbine, 177
Power corrections, 166–171
Power factors, typical, 1–2, 4, 6
Power of gas engines, 55
Power per cylinder in gas engines, 50
Power variation, gas turbine, 110
Precombustion chambers, 61, 67–69
Pressure corrections for gas turbines, 174
Pressure-loss corrections for gas turbines, 166–167, 169–170, 175
Pressure-volume diagrams for gas engines, 52–53
Projected life of a gas turbine, 120

Rankine cycle, 109–110, 157, 188–191
Rateau impeller, 217
Reaction, degree of, 226, 229
Rectified voltage commutator, 10–12
Resistance, 1
Richness, variation in, 63, 65–66
Rotor, 1

Index

Rubber couplings, 235
Rubber drive belts, 238

Sabath cycle, 16, 22–23
Scavenging loop, gas engine, 53–55
Seals, steam turbine, 230–231
Shaft alignment, 236
Shaft lines for gas turbines, 143–144
Shims, 236
Silencers
   exhaust, 39, 48
   intake, 46, 48–49
Slippage, 7
SO_3, condensation of, 213
Spark advance, gas engine, 98–102
   effect of, 59
Spark plugs, 77–78
   shielded, 80
Specification curve, gas turbine, 162–163
Specific consumption, 17
   gas engine, 97
Specific heat, 20
Specific power, 92–93
Speed regulation
   gas engine, 96
   motor, 7, 9–11, 12
      cost of equipment for, 14
Speed variation with power, gas turbine, 165
   motor, 15
Speeds of free-turbine gas turbines, 144
Spur gears, 237
Squirrel-cage motor, 1
Star coupled phases, 2
Star gear, 238–239
Startup
   current for, 2–3, 5
   gas turbine, 166, 181
   gas turbine and centrifugal compressor, 205
Startup rheostat, 2
Startup rotors, 3
Startup transformer, 2
Static, 83–84
Static-free ignition cable, 83
Stator, 1
Steam drum, generation with, 192–196
Steam generation, 192–206
Steam generators, once-through without drums, 196–197
Steam injections, gas turbine, 156–157

Steam power system, 193
Steam relief valve, steam turbine, 232
Steam turbines, 188–232
   calculations, 214–230
   characteristics, 214–232
   choice of, 188
   condensing, 230
   efficiency, 188–189, 228
   flow control, 231–232
   friction coefficients, 220, 223–224, 226, 228
   impulse, 214, 217–226
   reaction, 214, 226–230
   torque, 221
   velocity triangles, 214–215, 219, 222, 227
   wide vanes, 229–230
   work at the shaft, 17
Sulfur oxide emissions, 28
Sun gear, 238–239
Supercharged gas engines, timing of, 51–52
Supercharging, 86–92
   gas engine, 53–55
   pressures, 87
   systems, 88–91
Superheated steam, 189, 193, 196–197, 199, 201–204, 209, 211
Surge limit, gas turbine, 163
Synchronized asynchronous motors, 5–6
Synchronous motors, 5–6, 12–15

Temperature corrections for gas turbines, 173
Temperature-energy profiles, steam generation, 196, 198, 202, 211
Temperature limits versus torque, 92
Thermal efficiency
   gas engine, 50, 56, 57
   gas turbine, 108, 140, 171–178
   overall, 17
   typical, 18
Thermodynamic efficiency, gas turbine, 114
Thermosyphoning vaporizers, 199
Timing
   four-cycle gas engine, 51
   two-cycle gas engine, 52
Top dead center, 51
Torque
   motor, 2
   steam turbine, 221

Transformer coil assembly, 77
Turbocharger, 87–90
Turbo-cooling, 89
Two-cycle gas engine, 51

Unburned hydrocarbons, 28

Variable frequency, 9
Variable-speed alternating current motors, 6–16
Velocity triangles, steam turbine, 214–215, 219, 222, 227

Voltage, rectified, 10–11
Volumetric ratio, gas engine, 56–57
Vortex combustion, 158

Water injections, gas turbine, 155, 157
Wide vanes, in steam turbines, 229–230
  work at the shaft, effective, 17
    gas turbine, 115
    steam turbine, 215
    maximum recoverable, 24–25
Working fluids, choice of, 190–192
Wound rotor, 1